TERATOLOGY *Principles and Techniques*

*The Lectures and Demonstrations
were given at
the First Workshop in Teratology,
University of Florida,
February 2–8, 1964.*

Participants

ROBERT L. BRENT

F. CLARKE FRASER

E. MARSHALL JOHNSON

HAROLD KALTER

DAVID A. KARNOFSKY

NORMAN W. KLEIN

M. LOIS MURPHY

MEREDITH N. RUNNER

DAPHNE G. TRASLER

JOSEF WARKANY

JAMES G. WILSON

TERATOLOGY

Principles and Techniques

Edited by JAMES G. WILSON and JOSEF WARKANY

THE UNIVERSITY OF CHICAGO PRESS
CHICAGO AND LONDON

*A compendium of lectures and demonstrations from the First Workshop
in Teratology, held February 2–8, 1964, at the College of Medicine
of the University of Florida. The Workshop was sponsored by the Commission
on Drug Safety and was made possible by a supporting grant
from the Pharmaceutical Manufacturers Association.*

Library of Congress Catalog Card Number: 65-14432

THE UNIVERSITY OF CHICAGO PRESS, CHICAGO & LONDON
The University of Toronto Press, Toronto 5, Canada

FOREWORD

The Commission on Drug Safety, composed of fourteen distinguished scientists, came into being in August, 1962, following a grant from the Pharmaceutical Manufacturers Association. Lowell T. Coggeshall, Vice-President of the University of Chicago, formed the Commission and served as its Chairman. The Commission's charter called upon it to "seek new knowledge of the predictability of action of drugs in man," and the Commission's subsequent study and activity was shaped by this injunction.

The main objective of the Commission was development of a survey report on drug safety problems. This final report was published in June, 1964. It is comprised of several sections, one of which includes the individual reports of the seventeen subcommittees which the Commission had established to treat the major aspects of drug safety.

One of the seventeen subcommittees established by the Commission was that on Teratology, under the Chairmanship of Josef Warkany, Professor of Research Pediatrics, Children's Hospital Research Foundation, The University of Cincinnati. To develop the report of that Subcommittee, Dr. Warkany convened a "Conference on Prenatal Effects of Drugs" in Chicago on March 29–30, 1963. The proceedings of that Conference were published by the Commission in December, 1963.

The report of the Subcommittee on Teratology offered a number of recommendations, one of which was that workshops in teratology be undertaken. Teratologists could exchange information and techniques with scientists of related disciplines during these workshops.

Although the Commission on Drug Safety had earlier defined its role as advisory and recommending, rather than operational, when the report of the Subcommittee on Teratology was received, the need for immediate establishment of the suggested workshop was so patent that, by resolution of the Commission, planning for an initial "pilot" workshop was authorized, and a grant was given by the Pharmaceutical Manufacturers Association.

An invitation was extended to James G. Wilson, Professor and Chairman of the Department of Anatomy, University of Florida College of Medicine, to serve as Director of the Workshop and as host for this first program. Upon Dr. Wilson's acceptance of this invitation, planning was initiated in August, 1963, by Dr. Warkany and Dr. Wilson. The Workshop itself was held February 2–8, 1964, at the University of Florida. Forty-one participants, eighteen observers, and eleven faculty members took part. This volume presents the lectures and demonstrations.

In the initial planning for the Workshop, it was suggested that mimeographed summaries of lectures and demonstrations might be prepared for those who attended. This original concept has been modified in several ways. First, it became clear that a much wider audience could be served than the limited group which actually participated. Second, the materials themselves merited presentation in permanent form. Third, the Workshop served a definite need and similar programs should be continued in the future. This compendium may be of value for these future workshops. Finally, the scientists who took part in this first session called attention to the need for a permanent volume on procedures in teratology for day-to-day laboratory use.

JAMES G. WILSON
JOSEF WARKANY

CONTENTS

CHAPTER 1

DEVELOPMENT OF EXPERIMENTAL MAMMALIAN TERATOLOGY

JOSEF WARKANY

The uses of the past may be questionable in a book devoted to teaching of techniques and methods. Yet, by describing the development of experimental mammalian teratology I wish to show the justification and trends of this branch of teratology and the motives of some of the early workers in this field. Needless to say, this will be a very subjective presentation which may have little resemblance with the true history of our science; but the purpose of this admittedly incomplete presentation is to show the limited achievements of the past and to point out the vast needs and possibilities for the future.

When experimental mammalian teratology began in the 1930's, the science of experimental teratology practiced in lower classes of animals was at least 100 years old. Experiments on avian and amphibian eggs were performed most successfully during the nineteenth and our centuries, and many principles and fundamental facts were already known. But it was doubted by many investigators of human congenital malformations that the results obtained on eggs of lower classes could be applied to mammalian or human situations. It was thought that mammalian embryos and fetuses were so well protected by the maternal organism that they could not be modified by methods teratogenic to embryos of lower classes. It was thought that adverse environmental conditions would either kill a mammalian embryo or leave it unharmed. There were, of course, some observations which suggested that gross mechanical factors such as trauma, hemorrhage, or amniotic anomalies could deform a fetus *in utero*, and there were observations, clinical and experimental, showing that large (therapeutic) doses of X-rays could result in congenital malformations of the embryo (1). But such observations were considered as rare and exceptional and completely insufficient to explain the bulk of human and mammalian congenital defects. It had long been known that many congenital malformations were hereditary and passed on from generation to generation. When it was realized that genetic mechanisms can

Professor of Research Pediatrics, Children's Hospital Research Foundation, The University of Cincinnati, Cincinnati, Ohio.

be at work without clear transmission from parent to child, congenital malformations without signs of mechanical causation were often attributed to genetic factors, usually without proof. The terms "congenital" and "hereditary" were often used interchangeably. From the numerous statements made at that time, I quote only one as representative. In a textbook on human genetics (2), Fritz Lenz in 1931 stated: "When malformations are symmetrical, there can be little doubt that they are truly hereditary. If for instance, we find the same malformation in both hands, we can conclude with the greatest probability that the trouble is hereditary. The same thing applies to malformations which exhibit a serial homology, being of the same character both in the hands and in the feet."

The first experiments in mammalian teratology were done to test these assumptions. Methods more subtle than mechanical ones were sought to demonstrate environmental teratogenic effects upon the mammalian embryo.

The science of nutrition furnished the early tools for this task. Interest in the effects of maternal malnutrition upon the fetus had existed for some time, and in 1925 Jackson (3) summarized the earlier nutrition experiments which resulted either in normal development of the fetuses, reduced birth weight, or resorption, i.e., prenatal death. But congenital malformations were not effected by maternal inanition or general malnutrition. There was a single observation made by Zilva *et al.* (4) in 1921 during a study of the "fat-soluble factor" of four pigs with rudimentary extremities littered by a sow which was fed a deficient diet. It is clear that such a single observation could not be considered as proof of teratogenicity of maternal dietary deficiency. When Hale in 1932 made his first experiments with a deficient maternal ration, the science of nutrition had progressed further and the fat-soluble factor had been divided into vitamin A and vitamin D. Between 1933 and 1937 Hale reported that sows fed a vitamin A-free diet during early pregnancy farrowed pigs "all of which were born without eyeballs." Some of the pigs also had other defects such as facial clefts, accessory ears, misplaced kidneys, and occasionally malformed hind legs. Hale believed "the condition as illustrative of the marked effect that a deficiency may have in the disturbance of the internal factors that control the mechanism of development" (5). He also furnished good evidence that the malformations were not hereditary (6) and that normal pigs resulted from matings of a blind male to his dam and to his blind sister when the females were fed adequate rations (7). Although we consider now the pioneer experiments of Hale as well establishd, doubts were then expressed about the dietary etiology of the ocular changes (7), and it took a long time before these experiments were confirmed in pigs (8).

Although the size of such large animals could be advantageous sometimes, for instance, in experimental surgery, expenses and requirements of space preclude their use on a large scale. Further development of experimental mammalian teratology became possible when it was found that small rodents could also be used for this purpose.

When Dr. Rose Nelson and I began our experiments we were not acquainted with the brief and scattered reports of Hale. It was our idea to induce an endocrine deficiency in female rats to damage their young *in utero*. That this could be possible was suggested by the existence of endemic goiter and cretinism in many areas of the world where iodine deficiency prevailed (9). It was thought that maternal goiter of a severe degree led to cretinism of the child. To produce goiter we used the rachitogenic Steenbock and Black Diet No. 2965 which with or without vitamin D supplementation is goitrogenic. Animals fed this diet matured slowly, and breeding was delayed for many months. However, when the young were finally born, severe congenital malformations could be recognized in about one-third of them (10). We learned soon that we had failed to produce endemic cretinism; iodized salt added to the maternal diet prevented goiter in the mother and the young, but not the malformations; yet we had found a diet that produced skeletal malformations. When dried liver was given as a supplement, the malformations were prevented. Even before we knew what the preventive nutrient in liver was, various teratologic studies could be done. One could ascertain a definite pattern of skeletal malformations by clearing the newborn animals; one could study the histologic picture of the affected bones and follow the abnormal development back to mesenchymal stages of the skeleton; and one could demonstrate that liver was preventive up to the thirteenth day of gestation but not later. In fact, one could conclude (9) that the chronic nutritional deficiency of the mother, which began long before pregnancy and lasted throughout gestation, exercised its teratogenic effect upon the young between the thirteenth and fourteenth day of gestation. Although one could establish a critical period at which the abnormal environment brought on the malformations, nothing unusual happened to the mother at this critical period; the external situation was the same weeks before and weeks after the day on which the deforming changes occurred. These findings can still serve as a warning to those who believe that they can ascertain and time environmental teratogenic events with the help of an embryological timetable. But the most important result of these experiments was probably that even by 1940 one could point out that symmetrical, serial, and "familial" (repetitive) congenital malformations could be induced by an environmental manipulation.

In 1943 Miss Schraffenberger and I could show, thanks to the rapid progress of the science of nutrition, that the preventive factor in liver was riboflavin (11, 12), a rather startling finding in those days when skeletal malformations were often investigated by calcium- and phosphorus-balance studies. Subsequently purified diets became available, and it could be shown that such a semisynthetic diet without riboflavin induced skeletal defects identical with those induced by the original "goitrogenic" diet (11, 12).

With the advent of such purified diets the experiments could be varied by omission of various constituents of the diet, particularly of vitamins. There was an opportunity to induce vitamin A deficiency in pregnant rats and to conduct an experiment comparable to that of Hale. It was found that maternal vitamin A deficiency did indeed induce ocular anomalies in rat as in pig embryos (13), but in addition there were many other soft tissue malformations in the genito-urinary and cardiovascular systems and in the diaphragm (14). There were no malformations of the skeleton resembling those induced by riboflavin deficiency, and there was no overlapping of the two syndromes of congenital defects. The specificity of the defects produced by these dietary deficiencies is not known sufficiently, and one can still hear the opinion expressed that, of course, all bad maternal diets result in malformations, or that these deficiencies probably damage the placenta and affect the embryo secondarily. I have assumed all along that these deficiencies affect the embryo directly by withholding from him important nutrients, possibly enzyme constituents, and have not found any reason to change this working hypothesis; but I admit that direct proof for this assumption is still lacking.

Although the mechanisms of these deficiencies are still unexplored, the results were welcome finds for the teratologist interested in abnormal morphogenesis. The development of ocular malformations produced by vitamin A deficiency (colobomas, persistence of the primary vitreus, and others) could be followed through various embryonic and fetal stages (13). With Wilson we could study the great variety of aortic arch anomalies and cardiac defects (15). With vitamin A deficiency one could also throw into disorder the development of the genital ducts and produce male and female pseudo-hermaphroditism, a fact that should be of interest to the endocrinologist (16). Another interesting experiment resulted in a change of the vitamin A deficiency syndrome by restoration of vitamin A at various times during gestation (17). Thus a relationship could be established between the lack of these substances and the origin of malformations of various organs.

In 1948 two positive teratogens, nitrogen mustard and trypan blue, were found. Haskin (18) described severe malformations in the young of rats treated with a radiomimetic and cytotoxic nitrogen mustard, and comparable results were later also obtained in mice (19). Gillman *et al.* (20) introduced trypan blue as a teratogen which produced hydrocephalus, spina bifida, and other congenital anomalies in the young of treated rats. Trypan blue has been used by many investigators and still is today a favorite teratogen; but its mode of action is still unexplained. Gillman *et al.* believed that reproductive failure among Africans was due to chronic malnutrition which "could be complicated by the steady flooding of the circulation with particles derived from the abnormal metabolism or from the increased permeability of the gut, one of the consequences of malnutrition." They decided to employ "selected particles injected into animals in different physiological states." They chose trypan blue for this purpose and succeeded in producing a wide range of gross malformations in the young born to rats treated with dye before and during pregnancy.

In 1950 and 1952 Ingalls *et al.* (21, 22) reported on anoxia as a cause of fetal death and congenital defects in mice. These investigators placed female mice in a low pressure chamber for 5 hours at 260 to 280 mm Hg from the second to the eighteenth day of gestation. A significant number of young had hemivertebrae, fused ribs, cleft palate, and cranioschisis. Later Ingalls *et al.* (23) showed that hypoxia-induced congenital anomalies depended to a certain extent on the strain of mice employed in the experiment. Murakami *et al.* (24) examined embryos of female mice exposed to reduced atmospheric pressures long before birth and found changes in the central nervous system and eye. Degenhardt (25) extended this line of investigation to rabbits and found vertebral, costal, and ocular defects. Some of the malformed females were mated to malformed brothers, and all the young were normal (26).

Many hormones have been used as teratogens (27). Only a few can be mentioned here. In 1950 Baxter and Fraser (28) announced the production of congenital defects in the offspring of female mice treated with cortisone, and in 1951 Fraser and Fainstat (29) showed that the frequency of cleft palate produced in the offspring varied with the dose administered, the time of treatment, and the mouse strain used. Kalter and Fraser (30, 31) continued this experimental system in many studies which demonstrated the inheritance of susceptibility to the teratogenic action of cortisone. Needless to say, the blending of teratogenic and genetic studies was an important milestone in mammalian experimental teratology. Since Dr. Fraser's and Dr. Kalter's investigations appear in this book, I do not want to describe their work any further. Their experiments explain in part why those who work with hybrid animals cannot expect uniform results; but they show also that in addition to known genetic and exogenous teratogenic factors, some intangible influences may determine whether a malformation occurs or not. The uncertainties in well-controlled experiments should be made known to those who look for simple and general explanations of human congenital malformations.

The great interest in radiobiology that followed World War II led to renewed interest in X-rays as a teratogenic tool. We were curious to see whether rat embryos exposed to X-rays between the thirteenth and fourteenth days of gestation would produce malformations similar to those induced by riboflavin deficiency. The syndrome obtained was indeed superficially similar to that of the dietary deficiency, but there were also differences which have been described (32). Since 1950 many exact and detailed studies were made by Russell (33–35), Wilson (36–38), Hicks (39–41), and many others who successfully used X-rays as an acute and well-defined teratogen. This line of investigation, which is still being continued in many laboratories, represents a special chapter which I cannot treat adequately here. Review articles on this subject are available (42, 43).

I mentioned before that the use of purified diets made possible the testing of various vitamin deficiencies for their teratologic effects. It was found that under certain experimental conditions, maternal deficiency of vitamin E (44), pantothenic acid (45–47), folic acid, or nicotinic acid can result in congenital malformations of the young. To produce some of these deficiencies, special measures had to be taken to overcome the synthetic capacities of the maternal intestinal flora. To obtain folic acid (pteroylglutamic acid) deficiency in rats, an antimetabolite (48, 49) or an antibiotic (50) had to be introduced in the deficient diet. With the help of x-methyl folic acid (51), acute folic acid deficiency could be induced. This deficiency proved highly teratogenic in pregnant rats. The antimetabolites pantoyltaurine (46) or sodium *omega*-methyl pantothenate (47) were used to produce rapid pantothenic acid deficiency, and galactoflavin accelerated the effects of a riboflavin-deficient diet (52). The late Dr. Marjorie M. Nelson was a leader in this important development (53). Antimetabolites became powerful teratogenic tools which made possible conversion of long-term nutritional experiments into short-term chemical experiments. Whereas the chronic deficiency created an abnormal environmental background for the developing embryo which led to malformations at times of increased nutritional requirements, the metabolic antagonist created a sudden deficiency at selected stages of development. It is understandable that the new methods shortened teratologic experiments greatly; but the new sharp tools also made possible the production of large numbers of certain congenital malformations which previously had seldom been obtained. To give one example: In contrast to cleft palate, cleft lip is only rarely produced in rat fetuses by teratogenic procedures. But Nelson *et al.* (54) obtained cleft lip in more than 90 per cent of the young of rats subjected, with the help of an antimetabolite, to a transitory folic acid deficiency during days 9 to 11 of gestation. Being sharp tools, antagonists are often more effective and sometimes more devastating than simple deficiencies. Although galactoflavin added to a riboflavin-deficient diet induces skeletal malformations similar to those produced by the chronic dietary deficiency alone, it results also in additional cardiovascular, cerebral, and urogenital anomalies.

The nicotinamide antagonist, 6-aminonicotinamide, can inhibit the action of the vitamin in rat and mouse embryos, even without dietary restriction of the mothers (55, 56). This brings us to the long list of teratogenic chemical compounds found during the past 10 years which permit production of congenital malformations and monstrosities in fetuses after a single dose administered to a pregnant rodent. Many of these chemicals were at first used in tumor-inhibition screening programs and subsequently tested for teratogenic effects. Murphy (57) reviewed this topic and presented comparative dosages producing toxic and teratogenic effects on the pregnant rat and its fetuses for eight tumor-inhibiting chemicals. These compounds were alkylating agents, antimetabolites, and triazene. In addition to tumor-inhibiting compounds, many other drugs were found to be teratogenic in animal experiments. One of these is vitamin A which, given in excess to rats, mice, or rabbits (58–61), has become a favorite teratogen used in many experiments dealing with abnormal morphogenesis (61–63). It is curious that vitamin A deficiency and hypervitaminosis A have been found effective in the production of congenital malformations. Whether there is a common denominator or whether the two mechanisms have nothing in common remains unknown, since the mode of action has not been elucidated for either method. Some antibiotics (27, 65) and many hormones (27) have been found to be teratogenic in various experimental arrangements. Not all can be enumerated here; there are review articles available (27, 65, 66) to which the reader is referred. By 1961 it was known to the teratologist that many drugs, including salicylates (67), are teratogenic if given at the right time, in the right dose, to the right animal. It was also known that aminopterin (68–70) and certain progesterones (71) could be teratogenic in man. The surprising features of thalidomide (72) were that this drug which appeared completely harmless when given for a limited time in postnatal life had deleterious effects in minimal doses in prenatal life. It was also unusual that the rat embryo so extensively used in previous teratologic experiments proved very resistant to thalidomide.

As we review the field of experimental mammalian teratology we find that a huge mass of facts has been gathered within a short time by a few people. It is not surprising that there are loose ends and unfinished investigations everywhere. Some entire branches have failed to develop. Viral teratogens, for instance, clearly implicated in the etiology of human congenital malformations, have not been successfully used in mammalian experiments.

It is the purpose of this workshop to increase the number of investigators in this field. We need not only improved testing methods but also contributions to basic questions of teratology. I purposely pointed out that we failed to produce endemic cretinism but found something much better in the skeletal malformations induced by riboflavin deficiency. Serendipity is still our best ally. As hecatombs of rats are offered to an irate public, let us make use of these sacrifices to a rational purpose. The participants in this workshop should not only teach others the methods they learned but also transmit a desire for a better and deeper understanding of the phenomena observed.

There is also a moral obligation to communicate observations of interest to other investigators as is customary in all other fields of the modern scientific world. If these principles are adhered to, our workshop should prove a success.

REFERENCES

1. GOLDSTEIN, L. AND MURPHY, D. P. Etiology of ill health in children born after maternal pelvic irradiation: Defective children born after postconception pelvic irradiation. *Am. J. Roentgenol.* **22**: 322, 1929.
2. BAUR, E., FISCHER, E., AND LENZ, F. *Human heredity.* New York: The Macmillan Co., 1931, p. 286.
3. JACKSON, C. M. *Effects of inanition and malnutrition upon growth and structure.* Philadelphia: P. Blakiston's Son & Co., 1925.
4. ZILVA, S. S., GOLDING, J., BRUMMOND, J. C., AND COWARD, K. H. The relation of the fat-soluble factor to rickets and growth in pigs. *Biochem. J.* **15**: 427, 1921.
5. HALE, F. Pigs born without eyeballs. *J. Heredity* **24**: 105, 1933.
6. HALE, F. The relation of vitamin A to anophthalmos in pigs. *Am. J. Ophthal.* **18**: 1087, 1935.
7. HALE, F. The relation of maternal vitamin A deficiency to microphthalmia in pigs. *Texas J. Med.* **33**: 228, 1937.
8. PALLUDAN, B. The teratogenic effect of vitamin A deficiency in pigs. *Acta Veterinaria Scandinavica* **2**: 32, 1961.
9. WARKANY, J. Congenital malformations induced by maternal dietary deficiency: Experiments and their interpretation. *Harvey Lect.* **18**: 89, 1952–53.
10. WARKANY, J. AND NELSON, R. C. Skeletal abnormalities in offspring of rats reared on deficient diets. *Anat. Rec.* **79**: 83, 1941.
11. WARKANY, J. AND SCHRAFFENBERGER, E. Congenital malformations induced in rats by maternal nutritional deficiency. V. Effects of a purified diet lacking riboflavin. *Proc. Soc. Exper. Biol. Med.* **54**: 92, 1943.
12. WARKANY, J. AND SCHRAFFENBERGER, E. Congenital malformations induced in rats by maternal nutritional deficiency. VI. The preventive factor. *J. Nutrition* **27**: 477, 1944.
13. WARKANY, J. AND SCHRAFFENBERGER, E. Congenital malformations induced in rats by maternal vitamin A deficiency. I. Defects of the eye. *Arch. Ophthal.* **35**: 150, 1946.
14. WARKANY, J., ROTH, C. B., AND WILSON, J. G. Multiple congenital malformations: A consideration of etiologic factors. *Pediatrics* **1**: 462, 1948.
15. WILSON, J. G. AND WARKANY, J. Aortic arch and cardiac anomalies in the offspring of vitamin A deficient rats. *Am. J. Anat.* **85**: 113, 1949.
16. WILSON, J. G. AND WARKANY, J. Malformations in the genito-urinary tract induced by maternal vitamin A deficiency in the rat. *Am. J. Anat.* **83**: 357, 1948.
17. WILSON, J. G., ROTH, C. B., AND WARKANY, J. An analysis of the syndrome of malformations induced by maternal vitamin A deficiency. Effects of restoration of vitamin A at various times during gestation. *Am. J. Anat.* **92**: 189, 1953.
18. HASKIN, D. Some effects of nitrogen mustard on the development of external body form in the fetal rat. *Anat. Rec.* **102**: 493, 1948.
19. DANFORTH, C. H. AND CENTER, E. Nitrogen mustard as a teratogenic agent in the mouse. *Proc. Soc. Exper. Biol. Med.* **86**: 705, 1954.
20. GILLMAN, J., GILBERT, C., GILLMAN, T., AND SPENCE, I. A preliminary report on hydrocephalus, spina bifida, and other congenital anomalies in the rat produced by trypan blue. *S. African J. Med. Sci.* **13**: 47, 1948.
21. INGALLS, T. H., CURLEY, F. J., AND PRINDLE, R. A. Anoxia as a cause of fetal death and congenital defect in the mouse. *Am. J. Dis. Child.* **80**: 34, 1950.

22. Ingalls, T. H., Curley, F. J., and Prindle, R. A. Experimental production of congenital anomalies. Timing and degree of anoxia as factors causing fetal deaths and congenital anomalies in the mouse. *New Engl. J. Med.* **247**: 758, 1952.

23. Ingalls, T. H., Avis, F. R., Curley, F. J., and Temin, H. M. Genetic determination of hypoxia-induced congenital anomalies. *J. Hered.* **44**: 185, 1953.

24. Murakami, U., Kameyama, Y., and Kato, T. Effects of maternal anoxia upon the development of embryos. *Ann. Report,* Res. Inst. of Environmental Med., Nagoya Univ. 76, Mar., 1956.

25. Degenhardt, K. -H. Durch O$_2$-mangel induzierte Fehlbildungen der Axialgradienten bei Kaninchen. *Z. Naturforsch.* **96**: 530, 1954.

26. Degenhardt, K. -H. and Kladetzky, J. Wirbelsäulenmissbildung und Chordaanlage. Experimentelle teratogenetische und embryohistologische Untersuchungen bei Kaninchen. *Z. Mensch. Vererb.* **33**: 151, 1955.

27. Kalter, H. and Warkany, J. Experimental production of congenital malformations in mammals by metabolic procedures. *Physiol. Rev.* **39**: 69, 1959.

28. Baxter, H. and Fraser, F. C. Production of congenital defects in offspring of female mice treated with cortisone. *McGill Med. J.* **19**: 245, 1950.

29. Fraser, F. C. and Fainstat, T. D. Production of congenital defects in offspring of pregnant mice treated with cortisone. *Pediatrics* **8**: 527, 1951.

30. Kalter, H. The genetics and physiology of susceptibility to the teratogenic effects of cortisone in mice, Ph.D. Dissertation, McGill University, Montreal, 1953.

31. Fraser, F. C., Fainstat, T. D., and Kalter, H. Experimental production of congenital defects with particular reference to cleft palate. *Neo-Natales Studies* **2**: 43, 1953.

32. Warkany, J. and Schraffenberger, E. Congenital malformations induced in rats by roentgen rays. *Am. J. Roentgenol.* **57**: 455, 1947.

33. Russell, L. B. X-ray induced developmental abnormalities in the mouse and their use in the analysis of embryological patterns. I. External and gross visceral changes. *J. Exp. Zool.* **114**: 545, 1950.

34. Russell, L. B. X-ray induced developmental abnormalities in the mouse and their use in the analysis of embryological patterns. II. Abnormalities of the vertebral column and thorax. *J. Exp. Zool.* **131**: 329, 1956.

35. Russell, L. B. and Russell, W. L. An analysis of the changing radiation response of the developing mouse embryo. *J. Cell. Comp. Physiol.* Supp. 1, **43**: 103, 1954.

36. Wilson, J. G. and Karr, J. W. Effects of irradiation on embryonic development. I. X-rays on the tenth day of gestation in the rat. *Am. J. Anat.* **88**: 1, 1951.

37. Wilson, J. G., Jordan, H. C., and Brent, R. L. Effects of irradiation on embryonic development. II. X-rays on the ninth day of gestation in the rat. *Am. J. Anat.* **92**: 153, 1953.

38. Wilson, J. G., Brent, R. L., and Jordan, H. C. Neoplasia induced in rat embryos by roentgen irradiation. *Cancer Res.* **12**: 222, 1952.

39. Hicks, S. P. Some effects of ionizing radiation and metabolic inhibition on the developing mammalian nervous system. *J. Pediat.* **40**: 489, 1952.

40. Hicks, S. P. Developmental malformations produced by radiation. *Am. J. Roentgenol.* **69**: 272, 1953.

41. Hicks, S. P., *et al.* Mechanism of radiation anencephaly, anophthalmia, and pituitary anomalies. *Arch. Path.* **57**: 363, 1954.

42. Russell, L. B. The effects of radiation on mammalian prenatal development. In: *Radiation Biology,* edited by A. Hollaender. New York: McGraw-Hill Book Company, Inc., 1954, Vol. 1, p. 861.

43. Wilson, J. G. Differentiation and the reaction of rat embryos to radiation. *J. Cell. Comp. Physiol.* Supp. 1, **43**: 11, 1954.

44. Cheng, D. W. and Thomas, B. H. Relationship of time of therapy to teratogeny in maternal avitaminosis E. *Proc. Iowa Acad Sci.* **60**: 290, 1953.

45. Lefebvres-Boisselot, J. Rôle tératogène de la déficience en acide pantothénique chez le rat. *Ann. Méd.* (Paris) **52**: 225, 1951.

46. Zunin, C. and Barrone, C. Embriopatie da carenza di acido pantotenico. *Acta Vitaminol.* **8**: 263, 1954.

47. Nelson, M. M., Wright, H. V. Baird, C. D. C., and Evans, H. M. Teratogenic effects of pantothenic acid deficiency in the rat. *J. Nutrition* **62**: 395, 1957.

48. Hogan, A. G., O'Dell, B. L., and Whitley, J. R. Maternal nutrition and hydrocephalus in newborn rats. *Proc. Soc. Exper. Biol. Med.* **74**: 293, 1950.

49. Evans, H. M., Nelson, M. M., and Asling, C. W. Multiple congenital abnormalities resulting from acute folic acid deficiency during gestation. *Science* **114**: 479, 1951.

50. Giroud, A. and Lefebvres-Boisselot, J. Influence tératogène de la carence en acide folique. *Compt. Rend. Soc. Biol.* **145**: 526, 1951.

51. Nelson, M. M., Asling, C. W., and Evans, H. M. Production of multiple congenital abnormalities in young by maternal pteroylglutamic acid deficiency during gestation. *J. Nutrition* **48**: 61, 1952.

52. Nelson, M. M., Baird, C. D. C., Wright, H. V., and Evans, H. M. Multiple congenital abnormalities in the rat resulting from riboflavin deficiency induced by the antimetabolite galactoflavin. *J. Nutrition* **58**: 125, 1956.

53. Nelson, M. M. Mammalian fetal development and antimetabolites. In: *Antimetabolites and cancer* (Amer. Assoc. Advance. Sci. Monograph), edited by E. P. Rhoads. Washington, D.C.: 1955, 107.

54. Nelson, M. M., Wright, H. V., Asling, C. W., and Evans, H. M. Multiple congenital abnormalities resulting from transitory deficiency of pteroylglutamic acid during gestation in the rat. *J. Nutrition* **56**: 349, 1955.

55. Murphy, M. L., Dagg, C. P., and Karnofsky, D. A. Comparison of teratogenic chemicals in the rat and chick embryo. *Pediatrics* **19**: 701, 1957.

56. Pinsky, L. and Fraser, F. C. Production of skeletal malformations in the offspring of pregnant mice treated with 6-aminonicotinamide. *Biol. Neonat.* **1**: 106, 1959.

57. Murphy, M. L. Teratogenic effects of tumour-inhibiting chemicals in the foetal rat. *Ciba Foundation Symposium on Congenital Malformations,* 1960, pp. 78–107.

58. Cohlan, S. Q. Excessive intake of vitamin A as a cause of congenital anomalies in the rat. *Science* **117**: 535, 1953.

59. Cohlan, S. Q. Congenital anomalies in the rat produced by excessive intake of vitamin A during pregnancy. *Pediatrics* **13**: 556, 1954.

60. Giroud, A. and Martinet, M. Malformations oculaires avec fibrose du vitré chez des embryons de lapin soumis à l'hypervitaminose A. *Bull. Soc. Ophtalmol. de France,* No. 3, 1–11, 1959.

61. Kalter, H. and Warkany, J. Experimental production of congenital malformations in strains of inbred mice by maternal treatment with hypervitaminosis A. *Am. J. Path.* **38**: 1, 1961.

62. Giroud, A. and Martinet, M. Morphogenèse de l'anencéphalie. *Arch. Anat. Micr.* **46**: 247, 1957.

63. Giroud, A., Martinet, M., and Roux, C. Malformations urinaires dans l'hypervitaminose A. *Compt. Rend. Soc. Biol.* **151**: 1811, 1957.

64. Tuchmann-Duplessis, H. and Mercier-Parot, L. The teratogenic action of the antibiotic actinomycin D. *Ciba Foundation Symposium on Congenital Malformations.* London: J. & A. Churchill, Ltd., 1960.

65. Wilson, J. G. Experimental studies on congenital malformations. *J. Chron. Dis.* **10**: 111, 1959.

66. Giroud, A. and Tuchmann-Duplessis, H. Malformations congénitales. Rôle des facteurs exogènes. *Pathologie-Biologie* **10**: 119, 1962.

67. Warkany, J. and Takacs, E. Experimental production of congenital malformations in rats by salicylate poisoning. *Am. J. Path.* **35**: 315, 1959.

68. THIERSCH. J. B. Therapeutic abortions with a folic acid antagonist. *Am. J. Obst.* **63**: 1298, 1952.
69. MELTZER, H. J. Congenital anomalies due to attempted abortion with 4-aminoptero-glutamic acid. *JAMA* **161**: 1253, 1956.
70. WARKANY, J., BEAUDRY, P. H., AND HORNSTEIN, S. Attempted abortion with 4-aminop-teroglutamic acid (aminopterin); malformations of the child. *Am. J. Dis. Child.* **97**: 274, 1960.
71. WILKINS, L., *et al.* Masculinization of the female fetus associated with administration of oral and intramuscular progestins during gestation: non-adrenal female pseudo-hermaphrodism. *J. Clin. Endocrin.* **18**: 559, 1958.
72. LENZ, W., PFEIFFER, R. A., KOSENOW, W., AND HAYMAN, D. J. Thalidomide and con-genital abnormalities. *Lancet* **1**: 45, 1962.

HISTOLOGICAL ANALYSIS OF MALFORMATIONS PRODUCED BY VARIOUS AGENTS

To the teratologist, specimens obtained experimentally are of interest be-cause of their morphological properties, particularly if the end results resem-ble human malformations, because they permit observation of embryologic and fetal stages leading to the malformations present at birth and because they sometimes make possible investigation of mechanisms of abnormal de-velopment.

In many instances observations of external malformations supply informa-tion and data satisfactory to the analyst; in others, dissection of abnormal specimens demonstrates internal changes not visible externally, and clearing of specimens (1, 2) may reveal skeletal details not discernible on most careful external scrutiny or gross dissection. Serial histologic sectioning, a laborious and expensive procedure, furnishes additional knowledge of minute anom-alies which cannot be obtained with more expedient methods.

Only a few examples will be cited to illustrate the value of histologic methods, with emphasis on the fact that these methods have not been used sufficiently so far and that the entire area of microscopic analysis of experi-mentally produced abnormalities invites greater attention.

Both riboflavin deficiency (3) and thalidomide injury (4) result in mal-formations of embryos with preference for the "pre-axial" parts of the distal appendicular skeleton (Plate 1). The vertebral column (with the exception of tail vertebrae) is almost completely spared. Thus the patterns resulting from these different teratogenic actions are rather similar. Although experi-mental work of riboflavin deficiency was done chiefly in rats and thalidomide experiments were done in rabbits, there existed the possibility that these dif-fering methods acted through similar mechanisms. It was of interest, there-fore, that the skeletal changes comparable on external inspection or in

Some of the work cited here was supported by grant HD-00502, National Institutes of Health, U.S. Public Health Service.

cleared specimens differed markedly in histologic sections. Plate 2*A* shows a rudimentary tibia of a newborn rat whose mother had been fed a riboflavin-deficient diet. The tibia is fairly well outlined as a cartilaginous structure with some ossification in the diaphysis. Ossification, however, is very irregular; the conversion of columnar cartilage into bone is obviously disturbed; there is no well-defined line of ossification (Plate 2*B*) as cartilaginous islands and tongues are intermingled with the newly formed osseous trabeculae (Plate 2*A*) (5).

In contrast, the rudimentary tibia of the rabbit fetus whose mother had

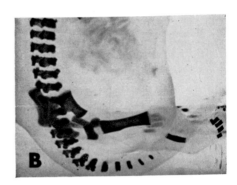
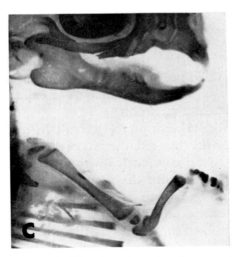

PLATE 1.—Cleared specimen (newborn rat, offspring of riboflavin-deficient mother). Minute radius (*A*) and absence of tibia (*B*).

Cleared specimens (rabbit fetuses; offspring of mothers treated with thalidomide). Absence of radius (*C*) and absence of tibia (*D*).

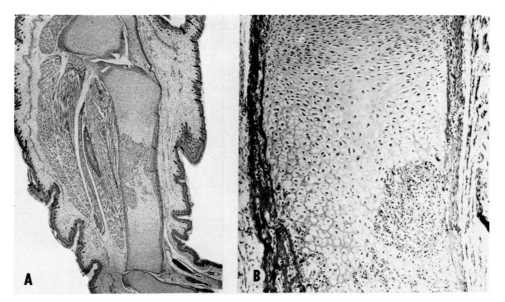

Plate 2.—Sections through tibia (newborn rat; riboflavin-deficient mother). Irregular and deficient ossification in both sections. *A* (approx. × 10); *B* (approx. × 70).

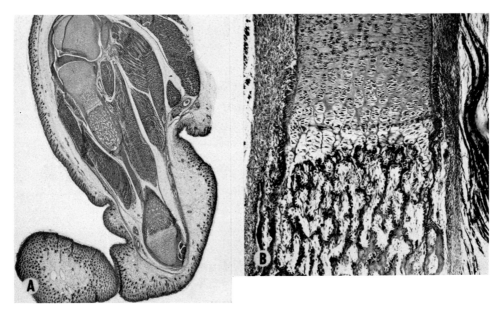

Plate 3.—Sections through tibia and lower end of fibula (rabbit fetus; mother treated with thalidomide).

   A, only proximal part of tibia present; distal part replaced by fibrous strand which inserts on distal end of fibula (approx. × 6).

   B, section through upper end of tibia shown in *A* (approx. × 60).

been treated with thalidomide is present only in its proximal part (Plate 3*A*). It continues as a fibrous band which inserts on the distal part of the fibula. Although this tibia is rudimentary, its proximal line of ossification is straight and regular, keeping cartilage and bone well separated (Plate 3*B*). The sections illustrated in Plate 4 represent lower legs which in cleared specimens would appear to have no tibiae. Yet in Plate 4*A* (riboflavin deficiency) there

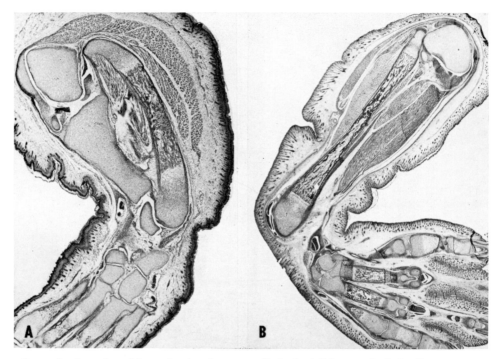

PLATE 4.—*A*, section of lower leg (newborn rat; riboflavin-deficient mother). Rudimentary cartilaginous tibia. The fibula which is much longer than the tibia shows good ossification (approx. × 12).

B, section through lower leg (rabbit fetus; mother treated with thalidomide). There is no tibia. Minute cartilage seen in angle between femur cartilage and fibula is a tibial rudiment (approx. × 5.5).

is a tibia formed in cartilage which would not be seen in the cleared specimen showing only bone. In Plate 4*B* (thalidomide) the tibia is actually absent except for minute cartilaginous rudiments on both epiphyseal ends. Thus by histologic examination, one gains the definite impression that the two teratogenic methods act through different mechanisms although both result in comparable skeletal malformations.

In our experiments with vitamin A deficiency, ocular anomalies were chosen as indication of "abnormality" of the young (6). Fibrous retrolenticular tissue was the most frequent ocular anomaly. Although some abnormal eyes could be recognized by external inspection, others that appeared externally normal were blind because there were fibrous tissue masses in place of the vitreous. Such fibrous remnants behind the lens could be identified only in histologic sections (Plate 5). Similarly aortic arch anomalies, minute cardiac defects (Plate 6), malformations of the genital ducts, and many other anomalies were discovered in serial histologic sections only. Such microscopic malformations could be missed without histologic examination. It is clear that incidence figures of abnormalities depend upon the methods of examination. For scanning of histologic serial sections, mounting on 35-mm film and covering with a plastic spray may be preferable to the conventional mounting on glass slides (7).

Histologic methods should be used particularly in studies of pathologic

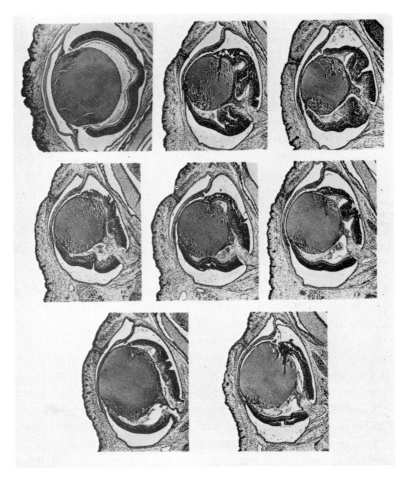

PLATE 5.—Section of eye on left in upper row is a control eye of newborn rat. All other sections (vitamin A deficient mother) show retrolental fibrous tissue (approx. × 16) .

15 *Histological Analysis of Malformations*

morphogenesis. Whereas a few years ago a well-preserved human or mammalian embryo with interesting malformations was a rarity considered to be worth intensive study, there are now available whole series of intact embryos and fetuses for observations of abnormal prenatal development. With the help of such series of experimentally produced malformations, many stages of abnormal development can be studied from early deviations through intermediate stages to end results at birth. Processes of degeneration and regeneration can be observed together and in sequence.

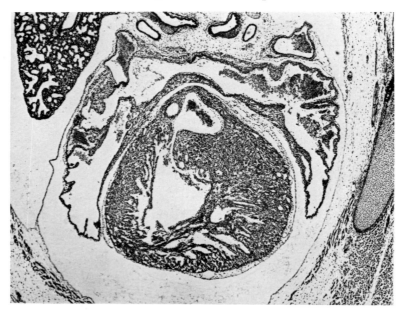

PLATE 6.—Section through heart of newborn rat (vitamin A deficient mother) showing minute ventricular septal defect (approx. × 35).

Development of spina bifida, for instance, can be demonstrated by following the fate of neural plates that fail to close. Plate 7A illustrates such a neural plate in an embryo obtained on the twelfth day of gestation. At this stage the neural plate should have been converted into a neural tube and spinal cord. Failure to close has grave consequences since the neural tissues remain on the body surface unprotected by the osseous, subcutaneous, and cutaneous covers present in normal fetuses. In Plate 7B the neural plate shows advanced development, lying on the dorsal body surface like an "open book." The anterior horns can be recognized in normal position, but the posterior horns appear overgrown and displaced laterad. Plate 8 illustrates the fate of such abnormal neural plates that persist to term. In Plate 8A the plate is still in good condition and well preserved, but it is without protection against the amniotic fluid and begins to degenerate in its lateral parts.

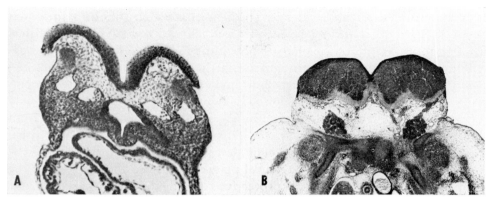

PLATE 7.—Sections of rat embryos (mothers treated with methyl salicylate [13]). *A*, open and overgrown neural plate of rat embryo, twelfth day (approx. × 65). *B*, open overgrown neural plate of rat embryo, sixteenth day (approx. × 21) .

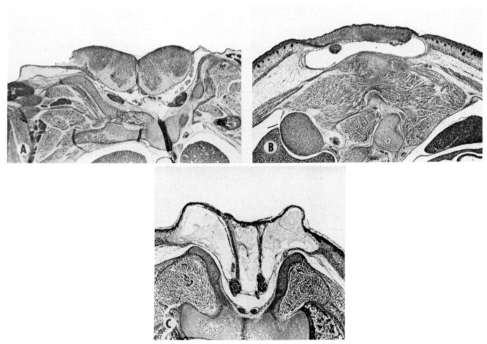

PLATE 8.—Sections through neural plates or their remnants of rat fetuses near term. *A*, mother treated with methyl salicylate. *B* and *C*, mothers treated with trypan blue (14) (approx. × 15).

Plate 8B shows an advanced pathologic stage in which the neural plate has lost its histologic structure and seems reduced to a homogeneous tissue mass. It is undermined on both sides by epithelial extensions of the skin which begin to separate it from its ventral pia-arachnoid cover. There is also a space forming ventral to the plate, the beginning of a fluid-filled "cyst" which will push the remaining neural tissue farther dorsad and away from the osseous trough formed by the vertebral bodies and processes. Plate 8C represents the final stage of degeneration. The neural plate has almost completely disappeared, but it is still indicated by the nerve roots which emanate from the dorsal cover of the cystic space. This cover consists chiefly of epithelium, but some nervous tissue remnants can be found occasionally at the origin of the nerve roots. As the neural plate degenerates and sloughs off, the epithelium assumes the role of a protective cover of the cystic space by growing over the wound left by the detached neural tissues. Plate 8C illustrates the principal features of a spina bifida cystica seen in transverse section. Thus one can follow in rat fetuses the development of a spina bifida from its beginning as a persisting neural plate through degenerative and regenerative processes to the final cystic form which is comparable to that seen in newborn children (8).

Similar processes take place if the neural plate remains open at the rostral end. At first one finds an exencephaly (Plate 9A). A well-developed but abnormal brain lies on the base of the skull in an untenable position. Like the neural plate in lower segments, the brain lacks the protective osseous and cutaneous covers. Degeneration sets in and hemorrhages occur within the brain substance (Plate 9B). The brain is lifted from the skull base (Plate 9C), and the nerves and blood vessels are unduly stretched. This leads to additional bleeding and destruction of brain tissue and to a state of anencephaly in which only few remnants of nervous tissue can be identified in a meshwork of meningeal and vascular debris lying on the cranial base (Plate 9D) (9–11).

It is obvious that such developmental deviations could be studied in great detail with new histochemical methods and would teach us a great deal about normal and abnormal embryology. Hydrocephaly and the Arnold-Chiari malformation as well as microcephaly and ocular, aural, cardiac, urogenital, and skeletal anomalies are obtained in teratologic experiments and await detailed analysis. Monstrosities, rare in human pathology, can be produced experimentally in rats or mice. The available specimens are in excellent states of preservation and can be sectioned serially, a feat seldom possible with the human counterparts because of their size. Geneticists have studied the pathologic embryology of many malformations arising in animals by mutation, but the wealth of material obtained in planned teratologic experiments has not been used sufficiently for that purpose (12).

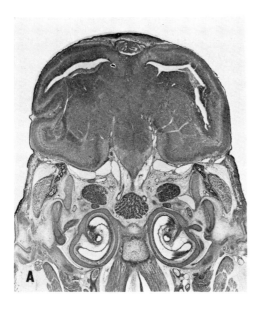
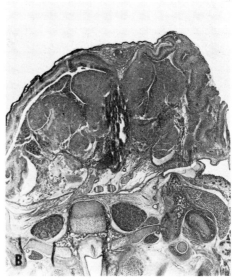
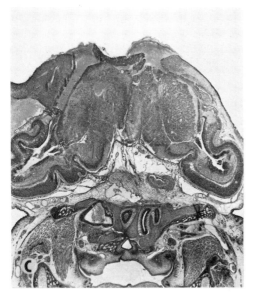
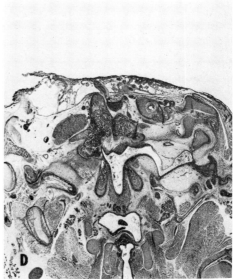

PLATE 9.—Sections through cranial region of rat fetuses near term. *A* and *B*, mothers treated with riboflavin-deficient diet and galactoflavin (15). *C* and *D*, mothers treated with large doses of vitamin A (16) (approx. × 12).

SELECTED BIBLIOGRAPHY

1. Dawson, A. B. Note on the staining of the skeleton of cleared specimens with alizarin red S. *Stain Technol.* **1:** 123, 1926.
2. Staples, R. E. and Schnell, V. L. Refinements in rapid clearing technic in the KOH-alizarin red S method for fetal bone. *Stain Technol.* **39:** 62, 1964.
3. Warkany, J. and Nelson, R. C. Skeletal abnormalities in the offspring of rats reared on deficient diets. *Anat. Rec.* **79:** 83, 1941.
4. Staples, R. E., Holtkamp, D. E., and Warkany, J. Effect of parental treatment with thalidomide on fetal development of rats and rabbits. *Abstr. Annual Meeting Teratology Society*, 1963.
5. Warkany, J. and Nelson, R. C. Skeletal abnormalities induced in rats by maternal nutritional deficiency. *Arch. Path.* **34:** 375, 1942.
6. Warkany, J. and Schraffenberger, E. Congenital malformations induced in rats by maternal vitamin A deficiency. I. Defects of the eye. *Arch. Ophthalmol.* **35:** 150, 1946.
7. Pickett, J. P. and Sommer, J. R. 35-mm Film as mounting base and plastic spray as cover glass for histologic sections. *Arch. Path.* **69:** 239, 1960.
8. Warkany, J., Wilson, J. G., and Geiger, J. F. Myeloschisis and myelomeningocele produced experimentally in the rat. *J. Comp. Neur.* **109:** 35, 1958.
9. Giroud, A. and Martinet, M. Hydramnios et anencéphalie. *Gynécol. Obst.* **54:** 391, 1955.
10. Giroud, A. and Martinet, M. Teratogenèse par hautes doses de vitamine A en fonction des stades du développement. *Arch. Anat. Microscop. Morphol. Expér.* **45:** 77, 1956.
11. Warkany, J. Experimental production of congenital malformations of the central nervous system. In: *Mental retardation*, edited by P. W. Bowman and H. V. Mautner. New York: Grune & Stratton, 1960.
12. Warkany, J. Pathology and experimental teratology. *Arch. Path.* **75:** 579, 1963.
13. Warkany, J. and Takacs, E. Experimental production of congenital malformations in rats by salicylate poisoning. *J. Path.* **35:** 315, 1959.
14. Gillman, J., Gilbert, C., and Gillman, T. A preliminary report on hydrocephalus, spina bifida, and other congenital anomalies in the rat produced by trypan blue. *S. Afr. J. Med. Sci.* **13:** 47, 1948.
15. Nelson, M. M., Baird, C. D. C., Wright, H. V., and Evans, H. M. Multiple congenital abnormalities in the rat resulting from riboflavin deficiency induced by the anti-metabolite galactoflavin. *J. Nutr.* **58:** 125, 1956.
16. Cohlan, S. Q. Congenital anomalies in the rat produced by excessive intake of vitamin A during pregnancy. *Pediatrics* **13:** 556, 1954.

CHAPTER 2

SOME GENETIC ASPECTS OF TERATOLOGY

F. CLARKE FRASER

The developing embryo can be likened to a ship being built. It begins as a set of plans—lines and words that constitute specific instructions as to exactly what the ship should be like. These lines and words correspond to the genes, which are specific instructions determining, by the sequence of nucleotides in the DNA, the structure of the proteins that make the developing organism what it is.

The plans for the ship are carried on pages of blueprints which correspond in the embryo to the chromosomes that carry the genes. Man has twenty-three pages of blueprints, in duplicate, one set coming from the mother and one set from the father.

Besides the instructions, to make a ship there must be building materials— wood, iron, canvas, etc.—which are incorporated into the ship according to the plans, and similarly the biological substrates—sugars, calcium, amino acids, etc.—are incorporated into the protoplasm according to the genetic instructions. At the risk of stretching the analogy too far, one may say that the instructions are translated into reality by the foremen (the messenger RNA) who read the plans and tell the workmen (ribosomes?) what to put where.

Finally the construction of the ship must take place in a not-too-unfavorable environment. The mammalian ship is built in a shed in which the building materials are stored and which protects it, more or less effectively, from wind and rain and usually from termites, rust, and rot.

Structural errors (malformations) in the ship may occur for a variety of reasons. First, there may be a *mistake in the plans*—a typographical error in the instructions so that some structure is put in the wrong place or made the wrong shape or not made at the right time or perhaps not made at all. These correspond to mutant genes which alter the structures or rates of synthesis of polypeptides in very specific ways. Theoretically a mutant gene should result in a specific biochemical defect, enzymatic or otherwise, and many such have been identified although they have not been associated so far, with anatomical malformations.

Second, there may be errors in the instructions resulting from an *extra*

Professor of Human Genetics, McGill University, Montreal, Quebec, Canada.

page of blueprint (or part of a page) being included in the plans or perhaps a page being left out. As would be expected, this leads to widespread confusion in the instructions, and abnormalities result in various parts of the ship. These errors correspond to the chromosomal aberrations, which characteristically lead to a wide variety of defects involving all the major organ systems.

Third, there may be errors due to the effects of *environmental agents*—the termites, rust, and rot—which correspond to the environmental teratogens. Of course the effects of the termites and rust will vary somewhat, depending on the quality of the building materials and how well they are put together, i.e., the "genetic background." This also applies to the effects of specific errors in the instructions.

Finally we come to a class of defects which result not from any one of the above major factors but from "a lot of little things"—the *multifactorial* group. Perhaps the wood is a little green, the iron a little rusty, the sails not quite of the proper strength, and the caulking material skimpy. No one of these things by itself would be noticeable, but acting together, and particularly in the presence of a few minor environmental stresses, the ship is found to be defective, either at launching or later, while being rigged, or later still after it has started on its life cruise. The multifactorial class of malformations, though they are the most difficult to analyze and to prevent, are also the most important since they are by far the most frequent class in man.

Errors in the instructions—mutant genes.—There is a very large number of mutant genes that cause congenital malformations of an almost infinite variety in both man and experimental animals. Those affecting the skeletal system have been reviewed recently by Gruneberg (1). According to current genetic theory the developmental effects of a mutant gene should be traceable to a specific change in the structure of, or rate of formation of, a particular protein. A number of such specific changes have been identified in hemoglobins, haptoglobins, gamma globulins, various enzymes, and other proteins (2), but so far no such change has been associated with an anatomical malformation. To discover the biochemical defect underlying a genetically determined malformation constitutes a great challenge, but so far we can only speculate as to where to look. Enzymes are an obvious place and, as our knowledge of fetal enzymology advances, surely enzymatic defects will be identified as the cause of certain gene-determined malformations. The task may not be easy, however, since the biochemical error may be a transitory one, present only at a critical stage of development. Morphologically, certain mutants look as if they resulted from errors in one or other of the *inductive processes,* and as these become better understood in biochemical terms (3) we may expect to find gene-determined alterations in them. Errors of *differentiation* may underlie malformations, but little is known of the biochemical basis of differentiation. The suggestion that allosteric proteins—in which the active site may be influenced by the presence of a biochemically unrelated substrate at another site on the same molecule—may play an important role in differentiation (4) implies also that genetically determined alterations in such proteins may change their developmental roles. The current interest in mechanisms that initiate or suppress gene activity (2) may also help to clarify problems of abnormal development.

Morphogenetic movements appear to depend, in part, on changes in the physicochemical states of cell membranes, and if one accepts that membrane structure is under genetic control, one might expect to find that some genetically determined malformations result from alterations of cell membrane characteristics. We must not forget, however, the increasing evidence that the cell membrane has "hereditary" characteristics which are not under nuclear control (5) and that the pattern of the cell cortex, which appears to be important for the early morphogenetic movements (6), may be self-replicating.

Pathogenetic studies of genetic malformations.—Not only does the existence of embryos that carry mutant genes causing malformations provide a chance to identify developmental errors in biochemical terms, but it allows one to describe how the malformation develops from the stage at which the first deviation from normal can be detected, and from this to infer something of the nature of the normal developmental processes. Although numerous examples of this approach could be cited, much work remains to be done.

For example, congenital hydrocephalus can occur for many different reasons, as shown by studies on gene-determined hydrocephalus in the mouse. One such gene, *hy-3*, produces its effect via a meningeal degeneration at the time of birth by blocking the cerebrospinal circulation, leading to an internal communicating hydrocephalus (7). In another, *hy-1*, there is widening of the fourth ventricle in the 12-day embryo, secondary stenosis of the aqueduct, and various other anatomical changes, particularly in the cerebellum (8), comparable to the Dandy-Walker syndrome in man (9). In a third, *ch*, the mutant gene affects the whole cartilaginous skeleton, but examination of early embryos shows that anomalies are present in the membranous skeleton before chondrification begins. Cartilage is normal, once formed, but onset of chondrification is delayed or prevented altogether, and there are both abnormal fusions and failures to fuse. Reduction in the mesenchymal basis of the skull leads to shortening of the chondrocranium which appears to force the normally growing hemispheres to bulge out dorsally with consequent compression of the lower parts of the brain, interference with cerebrospinal fluid flow in the region of the foramen of Magendie, and enlargement of the ventricles by hydrostatic pressure (10). These examples demonstrate how the same postnatal abnormality may arise in a variety of ways.

Examination of other parts of the skeleton reveals that all skeletal blastemata are involved, whether mesodermal or from the neural crest, and that they are involved whether they arise early or late in development. Thus the gene seems to affect a specific process, wherever and whenever it happens (11). Furthermore, it appears that chondrification does not set in until the blastema has reached a certain critical size—showing how knowledge of the abnormal can help in understanding the normal.

Many other mouse mutants also seem to affect the precartilaginous skeleton. The gene for "phocomelia," for instance, causes delay of precartilage formation and shortening of head, mandible, and limbs. Here, however, there are also differences in the distribution of cartilage-forming areas as well as enlargement of the footplates followed by pre-axial polydactyly (12).

A gene which produces effects more similar to those of thalidomide in man is dominant hemimelia (13) which causes reduction or absence of the tibia, pre-axial polydactylism or oligodactylism, syndactylism, kidney anomalies, absence of the spleen, intestinal atresias, and reduction or absence of bladder. Its developmental analysis is not yet complete, but identification of the biochemical defect might be very significant to the thalidomide problem.

Mutants affecting differentiation of the hands and feet are a fertile source of opportunities for studying developmental mechanisms. The mutant gene *luxate* (*lx*), for instance, causes a reduction or absence of the tibia, polydactyly or ectrodactyly, and horseshoe kidneys. There is a craniad shift in the position of the posterior limb buds at the time of their appearance. The hind limb buds are narrower than normal, and the apical ectodermal ridge (*AER*) is reduced. (The *AER* is a ridge of thickened ectoderm at the tip of the limb bud which appears to play an important role in differentiation of the distal limb.) Perhaps because of the narrowing, the tibial blastema is reduced or missing. This may be associated with missing pre-axial digits if the tibial blastema is missing. In other cases, the tibial blastema is narrowed and there is a relative excess of pre-axial mesenchyme distally with formation of extra digits (Plate 1). In luxoid (14) there are a caudal shift of the limb region and, in the forefoot, excess growth of pre-axial mesenchyme and an increase in length of the *AER*, leading to pre-axial polydactyly. In the hind foot, the mesenchyme and *AER* may either be increased or decreased with corresponding excesses or lacks of digit. The anomalies are present even before the mesenchymal blastema are formed. The gene for oligosyndactylism (*Os*) causes loss or fusion of digits. The shape of the footplate is abnormal as early as day 11; there is a reduction of material on the pre-axial border with an increase in the amount of cell death that occurs at this stage. When the digital blastemata are formed, digits 2 and 3 are closer together and more parallel than normal with varying degrees of fusion occurring subsequently (15).

Another mechanism occurs in embryos carrying the genes for syndactylism (*sm*) in which there is epidermal hyperplasia in various parts of the embryo, including the apical ectodermal ridge (1). The limb buds appear bloated with curved dorsal and concave ventral surfaces as if there were too much material in proportion to the circumference. Thus the digits do not spread out normally, and either soft tissue or bony fusions result (Plate 2). Again we have similar end results arising by different mechanisms.

Apart from the opportunity to analyze developmental mechanisms in mammals, these mutants are instructive in showing that a malformation commonly thought of as skeletal may in fact have its origin even before the

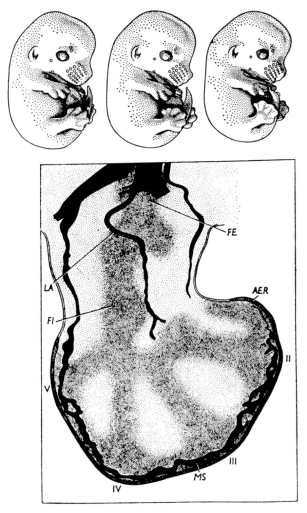

PLATE 1.—Effects of the luxate (*lx*) gene on the foot (from Gruneberg [1] after Carter [*J. Genetics*, Vol. 54, 1954]). *Top:* normal, polydactylous *lx/+* and hemimelic *lx/lx* embryos. *Bottom:* *lx/lx* footplate at $12\frac{1}{2}$ days. Note virtual absence of tibial blastema.

Genetic Aspects

mesenchyme begins to condense to form the precartilaginous anlagen of the skeleton.

Mutant genes producing cleft palate reveal a variety of mechanisms by which this malformation can arise. In urogenital homozygotes (*ur*) the palatine shelves are too narrow to meet even though they may move above the tongue at the right time (16; Plate 3). In phocomelic embryos there is

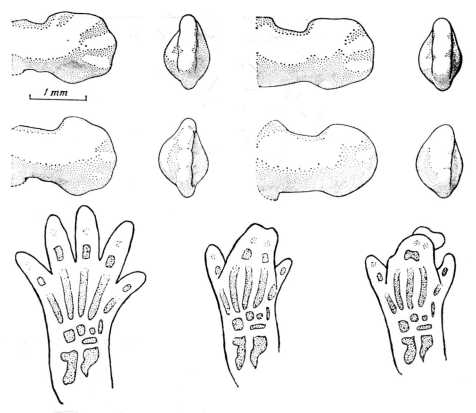

PLATE 2.—Effects of the syndactylism (*sm*) gene on the foot (from Gruneberg [1]). Right fore- and hind limbs of 12-day normal (*top*) and *sm/sm* (*middle row*) embryos. *Bottom row:* left hind-feet of normal (*left*) and *sm/sm* 15-day embryos.

an abnormal bar of cartilage in the area where palate shelf movement normally occurs (16). In shorthead, the tongue is short, and cupped between the front ends of the palatine shelves so it remains wedged between the shelves and prevents their closure (17; Plate 4). Again, it is clear that the same end result can be achieved in various ways.

There is a wealth of mutants involving the development of the central nervous system, the most interesting of which is a locus on chromosome IX of the mouse. A series of alleles is known at this locus which in various combinations affect all stages of development (18, 19). Homozygotes for the recessive lethal t^{12} become abnormal at the 30-cell stage at about 80 hours of age. RNA synthesis is reduced, and separation into embryonic disc and trophoblast never occurs. In t^0/t^0 animals, no mesoderm forms and death occurs before gastrulation, while t^4/t^4 embryos show abnormalities of the archenteron and die between day 7 and day 8. t^9/t^9 has duplications of the

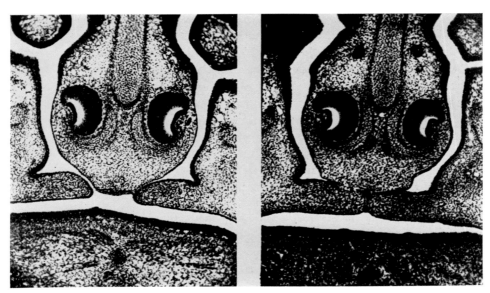

PLATE 3.—Cleft palate resulting from narrow palatal shelves in the *ur/ur* embryo (16). Normal on right.

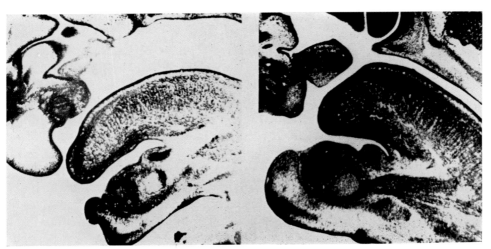

PLATE 4.—Cleft palate associated with short tongue and jaw in the *sh/sh* embryo (17). Normal on left.

27   *Genetic Aspects*

embryonic axis, and t^{w8}/t^{w8} shows pycnosis of the ectoderm and failure of the endoderm to differentiate. The dominant T mutant, the first to be described, results in absence of the posterior half of the body in the homozygote and in resorption of the distal tail and vertebral anomalies such as fusions or spina bifida in the heterozygote. These are preceded by gross abnormalities of the notochord which have been tentatively interpreted as stickiness by Gruneberg (20), who feels that the basic abnormality probably lies in the primitive streak. A mutant close to the T locus, *Kink,* causes complete or partial duplication of the embryonic axis, and another, probably allelic, causes irregular folding and branching of the neural tube and small or absent kidneys as in *Sd,* which will be mentioned later. Another group of alleles act late in development and result in pycnosis of the floor of the hindbrain and the ventral cord as well as hydrocephaly, microcephaly, and cleft palate, among other things. These and other mutants provide excellent material for studying the inductive mechanisms of the mammalian axial system.

I hope these few examples give some idea of the great variety of malformations produced by mutant genes. It is almost safe to say that every malformation produced by an environmental teratogen could be matched by one produced by a mutant gene.

The above examples take the approach of tracing back in embryogeny to the first visible deviation from normal and deducing from the nature of this deviation and those subsequent to it the developmental relationships concerned. Conclusions from this type of observation are useful as working hypotheses but they are not susceptible to direct proof by this approach.

Organ culture and mutant genes.—A more direct approach makes use of organ culture, and as techniques continue to improve, this method is likely to make important contributions to our knowledge of the genetic control of inductive and other developmental mechanisms.

A good example concerns the inductive relation of ectoderm and mesoderm in the developing limb bud of the chick. When the apical ectoderm ridge *(AER)* is removed from the early wing bud by EDTA, the distal limb does not develop, even if the missing ectoderm is replaced from elsewhere on the body. The *AER* transplanted to another part of the limb bud induces digits at the new site, so the *AER* seems to have an important inductive relationship to the underlying mesoderm. A new feature of the relationship is revealed when a mutant causing polydactyly is used. Polydactylous ectoderm combined with normal mesoderm and grown on the flank of the chick (not exactly organ culture, but almost) results in a normal limb; but polydactylous mesoderm with normal ectoderm results in a polydactylous limb (21). These and other experiments reveal a reciprocal interaction between ectoderm and mesoderm. A gene-controlled inductive influence from the mesoderm stimulates growth of the *AER,* which in turn stimulates growth and differentiation of the mesoderm.

Relatively few such experiments have been done in mammals, but I shall cite two. One involves an interesting mutant, *Sd*, which, besides affecting vertebral and skeletal development, causes anomalies of the genito-urinary system. Severely affected embryos have a persisting cloaca without rectum or anus, absence of kidneys, ureters, bladder, urethra, and genital papilla. Embryological study suggested that the effects on the ureter preceded those on the metanephrogenic blastema and that perhaps the kidney anomaly might depend on a lack of stimulus from the delayed uretral bud. Organ culture gives part of the answer. *Sd/Sd* kidney rudiment does have the capacity to differentiate into tubules, and a mutant ureter can induce tubule formation in mutant kidney mesenchyme. However, neither mutant kidney mesenchyme nor ureter differentiates entirely normally, and it is still not clear whether the mutant gene interferes with the timing of the inductive interaction or with the spatial relations (22). Nevertheless, organ culture has contributed information not obtainable by the descriptive approach.

The second example involves the *T* locus previously mentioned. In organ culture the differentiation of cartilage in somites of the 9-day mouse embryo depends on an inductive stimulus from the neural tube. Embryos homozygous for *T* show degenerative changes in notochord and neural tube which eventually result in absence of the posterior body. Neural tubes from *T/T* embryos, which are already grossly abnormal in structure, will, surprisingly, induce differentiation in the somites. On the other hand *T/T* somites, which are still normal in appearance, will not respond to the neural tube stimulus (23). Bennett speculates that perhaps the somites normally receive some influence from the notochord which is necessary before they respond to the neural tube inductive stimulus, and that this is lacking in *T/T* embryos. Obviously, inductive relationships are not as simple as had been assumed, and the use of mutants in this way should help to clarify the situation, even if at first it appears to complicate it.

Chromosomal anomalies.—The existence of autosomal trisomies in man, each producing a characteristic group of abnormalities, is now well recognized. It is just what one would expect from having an extra page in the plans. Why sex-chromosome aberrations do not seem to disturb development so severely is explained neatly (if not entirely) by the hypothesis that only one X is active in any given cell, no matter how many X's are present (24).

Mention should be made of the idea that chromosomal aberrations induced in somatic cells are the cause of malformations following exposure to teratogens. It is well known that radiation causes chromosomal breaks and rearrangements, but since these occur more or less at random from cell to cell it is difficult to see how they could cause the reproducible patterns of malformations that one sees following embryonic irradiation. Similarly, the chromosomal aberrations caused in mouse embryos following maternal treatment with 6-aminonicotinamide (25) are not likely to be the cause of the cleft palates and other malformations produced by the treament; it seems more probable that the general antimitotic effects of the drug might be involved in the teratogenic process.

Environmental teratogens.—Except for their interactions with genotype, these do not fall in the scope of this discussion.

The polyfactorial group.—Now we come to the most complicated group— those malformations determined by the interaction of many factors, both genetic and environmental. It is well known that the genotype of an organism can influence its response to an environmental agent. Sometimes the major variation results from a single gene difference, for example, the gene-determined deficiency of glucose-6-phosphate dehydrogenase in man that results in hemolysis following exposure to primaquine and other agents. This reaction may even happen prenatally, leading to a "congenital defect." However, the genetic variation in response to environmental factors is more often polygenically determined, and discussion of this question can logically begin with the subject of what Gruneberg (1, 11) has termed *quasi-continuous variation.*

If we take any biological characteristic such as height, weight, red cell diameter, time of menarche, etc., we find that it varies from individual to individual, so that there is a range of values with most individuals falling in the middle of the range and a few at either end. If we plot a frequency distribution, it is likely to follow a "normal" curve. If the character is markedly modified by a single gene difference with both alleles reasonably frequent in the population, the curve will be bimodal, but usually it is unimodal. This sort of curve will apply to the timing of developmental processes, among other things. Thus if we consider the embryonic age when the palate closes, or the optic cup reaches the ectoderm, or any other developmental event, it will be relatively early in a few animals and relatively late in others, even within the same litter. If we were able to observe the time of the event, say palate closure, we could plot a frequency distribution of times of palate closure relative to the age of the embryo. (If we use chronological age [see Plate 1 in Trasler's demonstration], the variability is increased, since embryos of the same chronological ages may have different developmental ages. It is more revealing to use the embryo's developmental age or "morphological rating" [see Plate 2 in Trasler's demonstration] as judged by various external features [26].)

Now suppose there is a threshold of some sort beyond which the process doesn't happen at all. Many examples of such thresholds are known. For instance, if the optic cup does not reach the ectoderm in time, the lens is not induced. The situation is similar for kidney induction by the ureter bud. If the palate shelves do not come up before a certain time, normal growth of the head carries them so far apart they cannot reach each other to fuse

(27). Another kind of threshold would be the "critical mass" of an organ anlage, below which differentiation does not occur. Such a threshold can break a continuous distribution of the type mentioned into discontinuous parts.

Gruneberg (1) has described a number of situations where this sort of thing occurs. One such is the absent 3d molar in the mouse, a defect which occurs with an 18 per cent frequency in the CBA strain and not in strain C57BL. A plot of frequency distribution of tooth size shows that in that

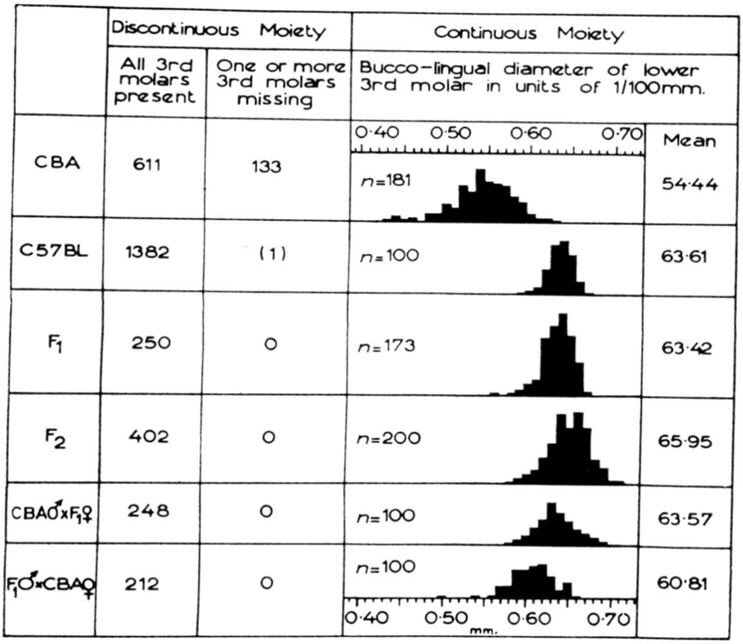

	Discontinuous Moiety		Continuous Moiety	
	All 3rd molars present	One or more 3rd molars missing	Bucco-lingual diameter of lower 3rd molar in units of 1/100mm.	
CBA	611	133	n=181	Mean 54·44
C57BL	1382	(1)	n=100	63·61
F₁	250	0	n=173	63·42
F₂	402	0	n=200	65·95
CBA♂×F₁♀	248	0	n=100	63·57
F₁♂×CBA♀	212	0	n=100	60·81

Fig. 1.—Relation of size of 3d molar to frequency of missing 3d molar in the mouse (from Gruneberg [1]).

strain which has affected animals, the average tooth size is small (Fig. 1). The frequency increases with parity and tooth size concurrently decreases. Furthermore, if a mouse has one 3d molar missing, the other is smaller than average. Maternal factors also seem important, but they may be transient; i.e., they vary from litter to litter, producing "clustering" as one often sees with environmental teratogens (Table 1). In short, it appears that there is a distribution of tooth sizes, multifactorially determined, and below a certain minimal size no tooth at all is formed so the continuous distribution of tooth sizes is broken into two discontinuous parts—tooth vs. no tooth. Various factors, genetic and environmental, can shift the distribution one way or the other.

Genetic Aspects

In the case of the palate, the discontinuity involves open vs. closed. If we plot, for the palate, a frequency distribution of the times when the shelves move from their vertical position on either side of the tongue to their horizontal position above the tongue, we will have a continuous distribution, but if we observe only the end result at birth, we have an "either-or" situation, a quasi-continuous distribution. This concept can account for many features of normal and abnormal development. Let us assume that the normal shape and position of the distribution is determined by the interaction of the genotype with the multifactorial environment. We know that in the A/Jax strain, palate closure normally occurs later, *relative to the developmental age of the embryo,* than in the C57BL strain. This can be seen in Plate 2 of Trasler's demonstration and is represented by the hypothetical diagram in Figure 2. The difference is determined by the genes, both of the fetus

TABLE 1

INCIDENCE OF NORMAL (N) AND ABNORMAL (A) MICE IN SUCCESSIVE
LITTERS (a, b, c, . . .) OF THREE MATINGS OF CBA MICE

Mating	a		b		c		d		e		f		g		Total	
	N	A	N	A	N	A	N	A	N	A	N	A	N	A	N	A
CBA 47/48....	6	..	..	8	4	..	6	..	5	1	..	..	..	..	21	9
CBA 49/50....	6	1	3	3	..	8	5	..	1	2	4	..	..	..	19	14
CBA 51/52....	6	2	2	..	3	1	3	..	6	..	..	5	2	1	22	9

From Gruneberg, H., showing that proportion of abnormal (missing 3d molar) animals is either high or low in individual litters (clustering).

and of the mother, with the strain which closes its palate later giving the highest frequency of cortisone-induced cleft palate (28, reviewed by 29).

Now suppose any embryo in which the shelves come up after, say, MR 16 had been reached would have a cleft palate—the "threshold" in Figure 2. Suppose further that the pregnant mother was treated with a teratogen, e.g. cortisone, at a dose which delayed palate closure *relative to the development of the rest of the embryo.* A large strain difference in the frequency of cleft palate results—100 per cent in the A/Jax and 20 per cent in the C57BL strain (Fig. 2). Note, however, that unless the closure process is markedly delayed, not all the embryos will be affected, even though they grow in the same uterus, have the same genotype, and are exposed to the same teratogen. Furthermore, the fact that the C57BL strain closes its palate earlier than the A/Jax strain does not mean that it forms its eye sooner or its vertebral anlage, so the fact that a teratogen causes more cleft palates in A/Jax than C57 does not mean that it will cause higher frequencies of other malformations as well. The genetic differences are *organ specific* (30).

Secondly, a mutant gene might act to delay shelf closure, just as cortisone does. If it acted on an animal already "predisposed" to cleft palate, such as the A/Jax strain, it might produce cleft palate in all the animals. The same mutant gene in the C57BL strain might result in only a portion—say 20 per cent—of the animals being affected, and we would say that the gene had *reduced penetrance*. (Fig. 2—for "treated," read "mutant.")

Now consider a case where a teratogen shifts the distribution just to the point where only a few animals are affected—a minimal or practically subclinical dose. Another teratogen shifts the distribution by the same amount.

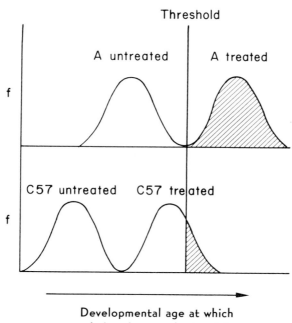

Fɪɢ. 2.—Hypothetical diagram showing time of palate shelf movement to the horizontal and threshold beyond which normal embryonic growth makes head too wide for shelves to meet, thus leading to cleft palate.

When the two act together, they might act additively, each shifting the distribution by the same amount. But this shifts the distribution to a point where the threshold falls in the steep part of the curve and results in a great increase in the proportion of animals falling beyond the threshold (Fig. 3). Thus, though the result as measured by cleft palate frequency appears to be *synergistic*, the two teratogens may actually be acting *additively*.

Inbred strains.—We have been referring to inbred strains from time to time, but have not said anything about their advantages and disadvantages. An inbred strain is formed usually by mating brothers with sisters for many generations. If the original stock is genetically heterogeneous, the result will

Genetic Aspects

be a stock homozygous at all loci. Thus if alleles *A* and *a* were segregating in the original stock, the inbred lines would very soon become homozygous, either *AA* or *aa*. Whether *A* or *a* became homozygous would be a matter of chance, unless there was deliberate selection. Some lines would become homozygous for *A* and others for *a*, and similarly for *B* and *b*, *C* and *c*, and any other genes that were segregating. The result would be a series of lines, all homozygous, but genetically different from one another. Thus, quite apart from their coat colors, the A/Jax and C57BL lines can be told apart by the shape of the embryonic snout, the structure of the scapula, etc. This does not mean, however, that inbred lines are phenotypically more constant, in spite of a widespread opinion to this effect. Homozygosity makes for

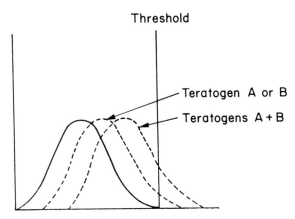

Fig. 3.—Diagram showing how two teratogens acting additively to shift distribution would appear to be synergistic as measured by cleft palate frequency.

genetic uniformity but is likely to increase somatic variability (30)—in other words, the shape of the distribution for a quantitative character may become flatter and wider. This variability may be one reason why inbred strains tend to have characteristic frequencies of "spontaneous" malformations—cleft lip and open eye in the A/Jax strain, microphthalmia and micrognathia in C57BL, and so on; i.e., the widened distribution overlaps a developmental threshold. If you want a minimum of variability and a maximum of vigor, use the F$_1$ of two inbred strains, as any corn geneticist will tell you.

Another point that should be emphasized is the matter of subline divergence. Mating brother to sister means that the stock is continually branching into sublines; the breeder may discard most of these and breed from only one, but if two sublines are maintained, for instance when the line is sent to another laboratory, they will become more and more distantly related to each other. One might think that if the line was homozygous to begin with, this would not matter. But there may be a certain amount of residual heterozygosity in the line, even after many generations, and there is also the fact of mutation, constantly creating genetic diversity, albeit at a very low rate. Thus sublines that have been separated for several generations may no longer be genetically identical, and a given strain today is not the same as it was 10 years ago (32). These facts have to be kept in mind when comparing results from one laboratory with those from another, or results over a period of years in the same line.

Inbred strains, in spite of their variability, have several advantages for teratological work. For one thing, they may be useful in studying the pathogenesis of a malformation. If a teratogen produces a malformation in only some of the animals in a litter, it may be difficult, when studying the embryological beginnings of the abnormal development, to tell which embryos would have been malformed at birth, since one usually has to kill the embryo to observe it. However, an inbred strain may be found which is susceptible to the effects of the teratogen on the organ being studied, so that the malformation can be produced in all the treated embryos, and one can be sure that any abnormalities noted in early development occurred in embryos that would have been malformed at birth. Thus, by studying the C57BL strain alone, we could not be sure that the cortisone-treated palate did not close and then open again, but since we never saw closed palates in A/Jax embryos when the treatment would have produced 100 per cent cleft palates, we could reject this possibility.

Inbred strain differences can be useful in another way. If one treats a number of strains with a teratogen and ranks them in order of susceptibility, any other teratogen that shows the same order of effect is likely to be acting by the same metabolic pathway. Such comparisons may give clues to mechanisms of action. Also, one can look for differences in the metabolic characteristics of the strains that show the same rank as the teratogenic susceptibilities and test the hypothesis that this difference is related to the susceptibility. A similar situation exists for species differences. Why is it impossible to produce cleft palates with cortisone in a rat, but easy in a mouse or rabbit? And why is thalidomide so teratogenic in man, but hardly at all in the mouse? An answer to these questions might be the answer to the nature of the drug's teratogenicity.

On the other hand, strain and species differences are also one of the main bugbears of teratology. One cannot extrapolate results from species to species, or even from strain to strain. This not only makes it difficult to compare results from laboratory to laboratory (since almost nobody seems to use the same drugs on the same kind of animals) but makes it impossible to predict from animal experiments whether a drug may be teratogenic in man. I have belabored this point elsewhere (33). Suffice it to say that a number of drugs in common medical use today are teratogenic in animals (though

Genetic Aspects

not detectably so in man), and at least one drug that is teratogenic in man is not much of a teratogen in several other animal species.

Interactions.—In conclusion, there should be some further mention of the fact that the embryo's genes interact with one another, with the mother's genes, and with environmental factors, often in very complicated ways. To begin with, the effects of major mutant genes may be modified greatly by the genetic background. For instance, the IX chromosome mutation *Fused* in the mouse was originally described as lethal in the homozygote with severe effects on both the axial and genito-urinary systems. The same gene is fully viable on other genetic backgrounds (34). Similarly, the effects of the Danforth short tail (*Sd*) mutant can be altered markedly by selection (Fig. 4). It is

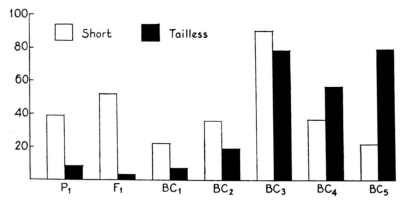

Fig. 4.—Effects of changing genetic "background" on expression of a mutant gene (from Gruneberg [1]). The proportion of mutant mice that are tailless increases in successive backcrosses to the inbred BALB/c strain.

not surprising to a geneticist, therefore, to find that the effects of an environmental teratogen are different on different genotypes (35).

Just how complicated gene-environmental actions can be is demonstrated by an experiment performed by Dr. Trasler and others in our laboratory, in which the effects of varying strain, dose of cortisone, maternal weight and diet on the frequency of cleft palate in the offspring were examined simultaneously (35).

It was found that each variable interacted with one or more of the others, so that one could not predict from the results of varying a factor in one strain how varying the same factor would affect another strain. For instance, doubling the dose had virtually no effect on cleft palate frequency in the C57BL strain on the Derwood diet, though it increased the frequency in all other combinations. Again, at the lower dose, the cleft palate frequency was higher with the Lab Chow than with the Derwood diet in the A/Jax strain, but lower in the C57BL strain. Furthermore, the cleft palate frequency decreased with increasing maternal weight in the A/Jax strain (more sharply on the Derwood diet) but did not vary significantly with weight in the C57BL strain. In short, these observations emphasize the fact that in teratology, as in many other fields, it is dangerous to generalize.

REFERENCES

1. GRUNEBERG, H. *The pathology of development.* Oxford: Blackwell's, 1963.
2. PARKER, W. C. AND BEARN, A. G. Application of genetic regulatory mechanisms to human genetics. *Am. J. Med.* **34**(5): 680, 1963.
3. HILLMAN, N. W. AND NIU, M. C. Chick cephalogenesis. I. The effect of RNA on early cephalic development. *Proc. Nat. Acad. Sci.* **50:** 468, 1963.
4. MONOD, J., CHANGEUX, J.-P., AND JACOB, F. Allosteric proteins and cellular control systems. *J. Molec. Biol.* **6**(4): 306, 1963.
5. SONNEBORN, T. M. Does preformed cell structure play an essential role in cell heredity? *The nature of biological diversity.* Edited by J. M. ALLEN, New York: McGraw-Hill Book Company, Inc., 1963.
6. CURTIS, A. S. G. The cell cortex. *Endeavour* **22:** 134, 1963.
7. BERRY, R. J. The inheritance and pathogenesis of hydrocephalus-3 in the mouse. *J. Path. Bact.* **81**(1): 157, 1961.
8. BONNEVIE, K. AND BRODAL, A. Hereditary hydrocephalus in the house mouse. IV. The development of the cerebellar anomalies during foetal life with notes on the normal development of the mouse cerebellum. I. Mat.-Naturv. Klasse, 1946, No. 4. *Skr. ut. av Det Norske Viden.-Akad.* (Oslo).
9. BRODAL, A. AND HAUGLIE-HANSSEN, E. Congenital hydrocephalus with defective development of the cerebellar vermis (Dandy-Walker syndrome). *J. Neurol. Neurosurg. Psychiat.* **22**(2): 99, 1959.
10. GRUNEBERG, H. Congenital hydrocephalus in the mouse, a case of spurious pleiotropism. *J. Gen.* **45:** 1, 1943.
11. GRUNEBERG, H. Genetical studies on the skeleton of the mouse. IV. Quasi-continuous variation. *J. Gen.* **51:** 95, 1952.
12. SISKEN, B. F. AND GLUECKSOHN-WAELSCH, S. A developmental study of the mutation "Phocomelia" in the mouse. *J. Exp. Zool.* **142** (1, 2, 3): 623, 1959.
13. SEARLE, A. G. The genetics and morphology of two "luxoid" mutations in the house mouse. *Genet. Res.,* in press (cited by Gruneberg [1]).
14. FORSTHOEFEL, P. The embryological development of the skeletal effects of the luxoid gene in the mouse, including its interactions with the luxate gene. *J. Morph.* **104:** 89, 1959.
15. GRUNEBERG, H. Genetical studies on the skeleton of the mouse. XXVII. The development of oligosyndactylism. *Genet. Res. Camb.* **2:** 33, 1961.
16. FITCH, N. An embryological analysis of two mutants in the house mouse both producing cleft palate. *J. Exp. Zool.* **136**(2): 329, 1957.
17. FITCH, N. Development of cleft palate in mice homozygous for the shorthead mutation. *J. Morph.* **109** (2): 151, 1961.
18. GLUECKSOHN-WAELSCH, S. Some genetic aspects of development. *Cold Spring Harbor Symp. Quant. Biol.* **19:** 41, 1954.
19. GLUECKSOHN-WAELSCH, S. Developmental genetics of mammals. *Am. J. Hum. Gen.* **13**(1, part 2): 113, 1961.
20. GRUNEBERG, H. Genetical studies on the skeleton of the mouse. XXIII. The development of brachyury and anury. *J. Emb. Exp. Morph.* **6**(3): 424, 1958.
21. ZWILLING, E. Genetic mechanism in limb development. *Cold Spring Harbor Symp. Quant. Biol.* **21:** 349, 1956.

22. GLUECKSOHN-WAELSCH, S. AND ROTA, T. R. Development in organ tissue culture of kidney rudiments from mutant mouse embryos. *Dev. Biol.* **7**: 432, 1963.
23. BENNETT, D. *In vitro* study of cartilage induction in *T/T* mice. *Nature* **181**: 1286, 1958.
24. LYON, M. F. Sex chromatin and gene action in the mammalian X-chromosome. *Am. J. Hum. Gen.* **14**: 135, 1962.
25. INGALLS, T. H., INGENITO, E. F., AND CURLEY, F. J. Acquired chromosomal anomalies induced in mice by injection of a teratogen in pregnancy. *Science* **141**: 810, 1963.
26. WALKER, B. E. AND FRASER, F. C. Closure of the secondary palate in three strains of mice. *J. Emb. Exp. Morph.* **4**:176, 1956.
27. WALKER, B. E. AND FRASER, F. C. The embryology of cortisone-induced cleft palate. *J. Emb. Exp. Morph.* **5**(2): 201, 1957.
28. TRASLER, D. G. AND FRASER, F. C. Factors underlying strain, reciprocal cross, and maternal weight differences in embryo susceptibility to cortisone-induced cleft palate in mice. *Proc. Tenth Int. Cong. Gen.* **2**: 296, 1958.
29. FRASER, F. C. The use of teratogens in the analysis of abnormal developmental mechanisms. *First Int. Conf. Cong. Malf.* Philadelphia: J. B. Lippincott, 1961, pp. 179–186.
30. GOLDSTEIN, M. B., PINSKY, M. F., AND FRASER, F. C. Genetically determined organ specific responses to the teratogenic action of 6-aminonicotinamide in the mouse. *Genet. Res. Camb.* **4**(2): 258, 1963.
31. MCLAREN, A. AND MICHIE, D. Variability of response in experimental animals. A comparison of the reactions of inbred, F₁, hybrid, and random-bred mice to a narcotic drug. *J. Gen.* **54**(3): 440, 1956.
32. GREWAL, M. D. The rate of genetic divergence of sublines of the C57BL strain of mice. *Genet. Res.* **3**:226, 1962.
33. FRASER, F. C. Experimental teratogenesis in relation to congenital malformations in man. In: *Proc. 2d Int. Conf. Cong. Malf.*, edited by M. FISHBEIN, in press.
34. DUNN, L. C. AND GLUECKSOHN-WAELSCH, S. A genetical study of the mutation "fused" in the house mouse, with evidence concerning its allelism with a similar mutation "kink." *J. Gen.* **52**(2): 383, 1954.
35. DAGG, C. P. The interaction of environmental stimuli and inherited susceptibility to congenital deformity. *Am. Zool.* **3**: 222, 1963.
36. WARBURTON, W., TRASLER, D. G., NAYLOR, A., MILLER, J. R., AND FRASER, F. C. Pitfalls in tests for teratogenicity. *Lancet* **2**: 1116, 1962.

STRAIN DIFFERENCES IN SUSCEPTIBILITY TO TERATOGENESIS: SURVEY OF SPONTANEOUSLY OCCURRING MALFORMATIONS IN MICE

DAPHNE G. TRASLER

Pregnant mice of two strains (A/Jax and C57BL) were given a 2-hr nicotinamide deficiency. This deficiency was created by giving them 6-aminonicotinamide in a dose of 19 mg/kg (body weight on conception day) on day 13.5 of gestation followed 2 hours later by a protective dose of niacin at 7 mg/kg. These females were killed on day 18, just before term. Their

Research Associate, Human Genetics Sector, Department of Genetics, McGill University, Montreal, Quebec, Canada.

uteri, intact with embryos, were cut out and fixed in Bouins for 4 or more days and then stored in 70 per cent ethanol (Tables 1 and 2). The method for examining these embryos as well as those from freshly killed females is described below. When 6-aminonicotinamide treatment is given on day 9.5 of gestation, skeletal abnormalities are produced in the young. A representative sample of five litters from each strain with embryos cleared and stained for bone with alizarin red are on hand.

TABLE 1

EMBRYO ABNORMALITIES AFTER 2-HR NICOTINAMIDE DEFICIENCY ON GESTATION DAY 13
(FIXED *in Utero* ON DAY 18)

Strain	Total Litters	Total Embryos	Number with Cleft Palate	Number with Cleft Lip	Per Cent Cleft Palate	Embryos with Other Abnormalities
A/Jax......	40	268	194	18	72	
C57BL......	25	125	35	....	28	1 (small lower jaw) 1 (microphthalmos) 2 (hind legs short)

TABLE 2

EMBRYO ABNORMALITIES FROM A RE-EXAMINATION (BY D. G. T.) OF A
RANDOM SAMPLE OF THE LITTERS IN TABLE 1

Strain	Total Litters	Total Embryos	Number Cleft Palate		Number Cleft Lip		Embryos with Other Abnormalities
			D. G. T.	Participants	D. G. T.	Participants	
A/Jax.....	16	118	96	99	14	7	11 (open eyes) 1 (polydactyly) 30 (with microphthalmia or anophthalmia) 2 (small lower jaw) 2 (short snout) 2 (hind legs short)
C57BL....	16	92	30	34	....	....	

It appears that participants did not always distinguish between A/Jax embryos carrying the *spontaneous* anomaly of cleft lip (with or without cleft palate) and the nicotinamide deficiency-*induced* cleft palate alone. In Table 2 it can be seen that a number of other spontaneous abnormalities were missed by workshop participants—in A/Jax open eyes and polydactyly; in C57BL microphthalmia, anophthalmia, small lower jaw (micrognathia and microstoma), and short snout.

PROCEDURE

Place an A/Jax strain uterus in a petri dish, count the number of embryos and resorptions, and note their position. Check card to see that numbers correspond. With forceps, dissect out embryo No. 1. Examine toes, feet, tail, umbilical cord, genitalia, ears, eyes, and snout. If necessary, compare with control embryo. Examine palate by cutting into neck and cutting sides of mouth so the lower jaw with tongue can be retracted. Write on card the embryo number, closed palate or cleft palate, and any other abnormalities.

Do the same with each of the other embryos. Replace embryos and uterus in bottle. Then examine a C57BL strain uterus in the same way. Hand in results so that the total per cent cleft palate for the two strains may be calculated.

SKELETAL PREPARATIONS

The embryos from a treated litter are arranged in order in a petri dish. The embryo on top is a control embryo (placed at right angles to the others). Examine each embryo under the dissecting microscope; look at the limbs and especially at the vertebrae and ribs. Write down rib abnormalities (missing or fused) and which of the thirteen ribs are concerned. Note vertebral fusions and whether they are in the thoracic, lumbar, or sacral region. Hand in your results so that the total per cent of each type of abnormality can be calculated for each of the strains.

SELECTED BIBLIOGRAPHY

GOLDSTEIN, M. B., PINSKY, M. F., AND FRASER, F. C. Genetically determined organ specific responses to the teratogenic action of 6-aminonicotinamide in the mouse. *Genet. Res. Camb.* 4(2): 258, 1963.

PINSKY, L. AND FRASER, F. C. Production of skeletal malformations in the offspring of pregnant mice treated with 6-aminonicotinamide. *Biol. Neonat.* 1(2): 106, 1959.

SPONTANEOUS ABNORMALITIES

In the embryos examined in the C57BL and A/Jax mouse strains, list the abnormalities induced by 6-aminonicotinamide and then list the spontaneous abnormalities seen. Examine samples of other spontaneous abnormalities that have been found in these strains.

PROCEDURE FOR MORPHOLOGICAL RATINGS AND PALATE STAGES

Palate closure in the mouse takes place sometime between gestation day 14 and early in day 15. Look at illustrations and description of *stages of palate closure,* then look at Table 3. The number of embryos found at a given palate stage and chronological time in gestation have been plotted for strains A/Jax (*A*) and C57BL (*B*) and crosses between them. Note, for instance, that onset of palate closure is later in *A* than in *B*. Also note that chronological age is not a good measure of time of palate closure, as even within an inbred mother the embryos can vary widely in palate closure stage and stage of development.

Developmental stage is a better indicator of embryonic age than chronological time. Thus the amount of webbing of the toes, whether the eyes are open or shut, and other morphological characters are given values. The sum

Daphne G. Trasler   40

TABLE 3

PALATE STAGE VS. CHRONOLOGICAL RATING

Chronological Rating (day/hr)	A Palate Stage							A×B Palate Stage							B×A Palate Stage							B Palate Stage							AB×A Palate Stage							BA×A Palate Stage						
	1	2	3	4	5	6	7	1	2	3	4	5	6	7	1	2	3	4	5	6	7	1	2	3	4	5	6	7	1	2	3	4	5	6	7	1	2	3	4	5	6	7
13/8	6																																									
16	5														8														7							8						
20															8	1		1											5							19	1					
22															5							7							11			1		5	2	20						
14/0								6							10	1	1	2	2	4	5	19	1	1	2			1	12	1		1	1	14	3	2	1			1	7	5
2								9			4				7	2		3	3	5	5	10	1	2	3	2	1	1	14	1	4	1	1	11	1	4			1	1	2	5
4	1					6		8				1	2		9	2	1	1	2	8	2	7	2	1	3		2	3	1	1	1	2	1	10	4			1	1	7	7	5
6	3					4		11		2	2	1	2	7	1	1	3	3	10	2	3	3	3	2	1	1	9	1	4	1	2	1	1	9	3		1	2	2	7	8	
8			1			1		14	1	3	1	2	4	14	1		1	2	4	2	13	11	2	3	1	1	5	5	2			1	1	5	3			1	1	1	1	8
10						6		1			1		2	6		3		1	1	1	4						3									3		1		1	4	4
12	7			1	1	4	7							4	1				1	1	13	1		1	1		3	1	1					1	7					8	7	
14	9	1		1	1	1	3		1				1	10						2	6					1	7	3							8							9
16	4	3		1	1	4	7	2				1	2								4		1		1	1	4	1														
18	2	1		1	1	2	6	1																			2	1							5							
20	12	3	1	1		1	8	2		1			1	6												1	10	10														
22	1	1						3						8												1	2	2														
15/0						6																																				
2	1	1	2	1	1	1	1																																			
4																																										
6																																										

of these values is called the morphological rating. Look at illustrations and description of *morphological rating* (Table 4). Then look at Table 5. Note that the palate closure begins later and proceeds more quickly in *A* than in *B,* while *A* × *B* and *B* × *A* are intermediate.

Examine the embryos provided and morphologically rate them. Note that palate closure occurs earlier in relation to development in the C57BL strain than in the A/Jax strain.

TABLE 4

DEVELOPMENTAL FEATURES AND THEIR ARBITRARILY ASSIGNED VALUES USED
TO CALCULATE MORPHOLOGICAL RATINGS FOR MOUSE EMBRYOS

VALUE ASSIGNED TO EACH FEATURE	DEVELOPMENTAL FEATURES				
	Forefeet	Hindfeet	Ears	Hair Follicles	Eyes
−2	un. pl.				
−1	in. pl.	un. pl.	sm. ra.		
0	plate	in. pl.	ra.	none	open
1	$\frac{3}{4}$ webbed	plate	$\frac{1}{3}$ closed	body	$\frac{1}{2}$ closed
2	$\frac{1}{2}$ webbed	$\frac{3}{4}$ webbed	$\frac{1}{2}$ closed	b.f.s.	closed
3	$\frac{1}{4}$ webbed	$\frac{1}{2}$ webbed	$\frac{2}{3}$ closed	b.s.	
4	free	$\frac{1}{4}$ webbed	$\frac{3}{4}$ closed	b.h.	
5	$\frac{1}{4}$ fused	free	closed		
6	$\frac{1}{2}$ fused	$\frac{1}{4}$ fused			
7	$\frac{3}{4}$ fused	$\frac{1}{2}$ fused			
8		$\frac{3}{4}$ fused		w.b.	

Abbreviations:
un. pl., unindented plate
in. pl., indented plate
sm. ra., small right-angled pinna
ra., right-angled pinna

b.f.s., follicles on body and a few on side of head
b.s., on body and side of head
b.h., on body and head
w.b., wrinkled body

REFERENCES

1. WALKER, B. E. AND CRAIN, B. Effects of hypervitaminosis A on palate development in two strains of mice. *Amer. J. Anat.* **107**(1): 49, 1960.
2. GRUNEBERG, H. The development of some external features in mouse embryos. *J. Hered.* **34**:89, 1943.

SELECTED BIBLIOGRAPHY

CALLAS, G. AND WALKER, B. E. Palate morphogenesis in mouse embryos after X-irradiation. *Anat. Rec.* **145**(1): 61, 1961.

FRASER, F. C. The use of teratogens in the analysis of abnormal developmental mechanisms. *First Inter Conf. on Cong. Malformations.* Philadelphia: J. B. Lippincott Co., 1961, pp. 179–86.

TRASLER, D. G. AND FRASER, F. C. Role of the tongue in producing cleft palate in mice with spontaneous cleft lip. *Dev. Biol.* **6**:45, 1963.

TRASLER, D. G., CLARK, K. H., AND FRASER, F. C. No cleft palates in offspring of pregnant mice given cortisone after fetal palate closure. *J. Hered.* **47**(2): 99, 1956.

WALKER, B. E. AND CRAIN, B. Abnormal palate morphogenesis in mouse embryos induced by riboflavin deficiency. *Proc. Soc. Exper. Biol. Med.* **107**:404, 1961.

TABLE 5

PALATE STAGE VS. MORPHOLOGICAL RATING

Morphological Rating (days)	A Palate Stage							A×B Palate Stage							B×A Palate Stage							B Palate Stage							AB×A Palate Stage							BA×A Palate Stage						
	1	2	3	4	5	6	7	1	2	3	4	5	6	7	1	2	3	4	5	6	7	1	2	3	4	5	6	7	1	2	3	4	5	6	7	1	2	3	4	5	6	7
−3	12																																									
−2																																										
−1	1																																									
0	1							1							3							4																				
1	1							1							3							2															1						
2	3							2							7							6								4							2						
3	3							2							3							2								4							3						
4	2							22							12	1						38	3							16							16						
5	2							11							4							1	1	1						6	1						14						
6	9	3						6							2		1					1	1	1		5	5	3		6							3						
7	7	3	2	1		9	1	3	4						4	2		1	7	2	2	1	2	2	1	1	10	2	13	5	1	1		20	1	7		1				
8	8	3	4		1	8	6	1	6				7		4	3				10	8	2	2	5	2		8	6	5	1	4	1	3	12	6	10	3	1			3	1
9	3	1	6	3	1	5	10	1	1			7	11	1	3				1	11	14	1	1	1	5	1	3	8	2		4	1	9	9	5	3	1	1	1	2	7	4
10	2	1	1		1	11	1		1		2	1	7	4					1	10	7		1		1	2		9			1			9	3					1	7
11			1			7	6					7	7	6						4	13							6							3			1	2	11	7	4
12			1			4	10						4	16							14							9							1			3		6	4	7
13					1	5	12							12							13																					16
14							3							4							7							6													2	
15							6							6							1							2													6	
16							1							2							1							1													2	
17							1							4							2							3													3	
18							2							4														2														
19																																										
≥20																												1							1							

WALKER, B. E. AND FRASER, F. C. Closure of the secondary palate in three strains of mice. *J. Embryol. Exp. Morphol.* **4**: 176, 1956.

WALKER, B. E. AND FRASER, F. C. The embryology of cortisone-induced cleft palate. *J. Embryol. Exp. Morphol.* **5** (2): 201, 1957.

Relation of tongue and palatine shelves during normal secondary palate closure.—STAGE 0 (Plates 1–2). About 11 days after conception, the upper lip and primary palate are formed from the median nasal, lateral nasal, and maxillary processes. The shelves that will form the secondary palate appear as longitudinal ridges on the upper surface of the oral cavity. Just before palate closure starts, about 15 days after conception, the tongue is bounded laterally by the vertical palatine shelves, superiorly by the floor of the skull, inferiorly by the mandible, and anteriorly by the primary palate. The tongue is roughly rectangular in cross-section, being more than half as thick as it is wide, and appears to be compressed by the incurving shelves. When the tongue is removed, the shelves are found to lie well apart posteriorly as well as anteriorly.

STAGE 1 (Plates 3–4). Stage 1 begins when the shelves begin to move into their final position above the tongue. The tongue is still cupped between the shelves, but its tip has moved forward and under the primary palate. In the intact fixed embryo, the tongue tip can be seen protruding between the lips, while previously, at Stage 0, it could not. The shelves are not horizontal and above the tongue posteriorly, and their posterior ends are closer together than in Stage 0.

STAGE 2. The shelves continue to move over the tongue and toward the midline posteriorly, and Stage 2 is reached when the posterior half of the shelf has become horizontal above the tongue.

STAGE 3 (Plates 5–6). In Stage 3, one shelf attains a horizontal position above the tongue, which is now flattened on this side so that it is wider than before. The other shelf still remains vertical anteriorly and cups the other side of the tongue as before. The tongue is still closely applied to the shelves and follows their contours.

STAGE 4 (Plate 7). When both shelves have assumed a horizontal position and lie dorsal to the tongue, they are at first separated by a small space which is soon bridged by further flattening of the shelves.

STAGE 5. This is the stage at which fusion of shelf epithelium begins.

STAGES 6 (Plate 8) AND 7. Epithelial fusion spreads anteriorly and posteriorly (Stage 6) until the shelves are fused throughout their length (Stage 7).

Morphological rating.—Morphological rating of mouse embryos for days 13, 14, and 15 of gestation as described by Walker and Crain (1) is presented below. (A less detailed description of embryos from day 9 to day 18 of gestation can be found in Gruneberg [2]).

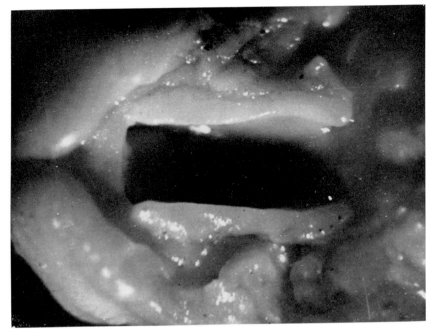

PLATE 2.—Stage 0, tongue removed

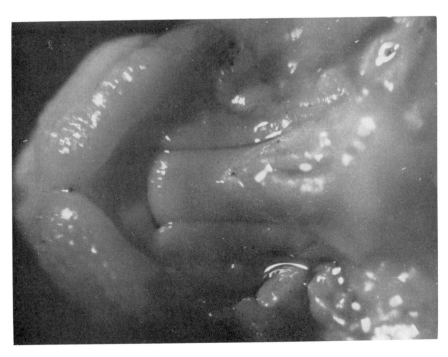

PLATE 1.—Stage 0, tongue in place

45

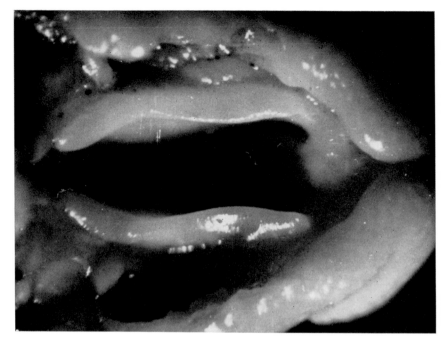

PLATE 4.—Stage 1, tongue removed

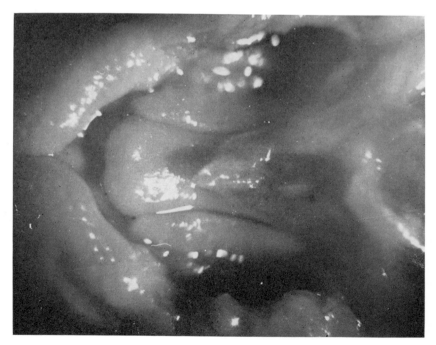

PLATE 3.—Stage 1, tongue in place

46

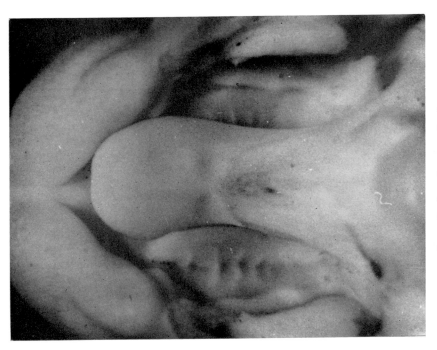

PLATE 5.—Stage 3, tongue in place

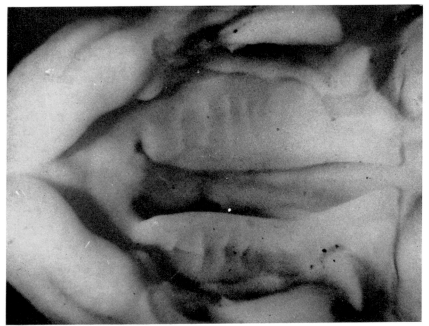

PLATE 6.—Stage 3, tongue removed

47

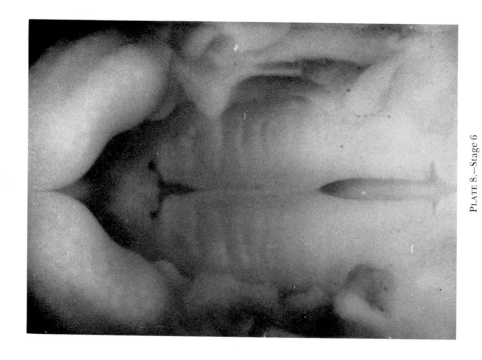

PLATE 8.—Stage 6

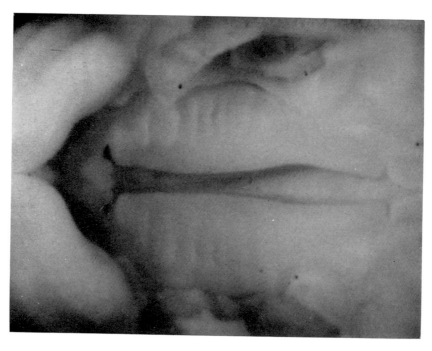

PLATE 7.—Stage 4

48

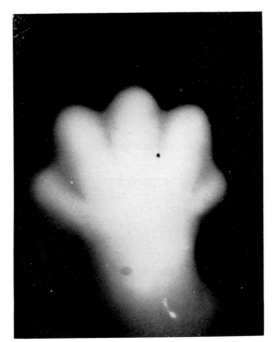

PLATE 9.—Plate

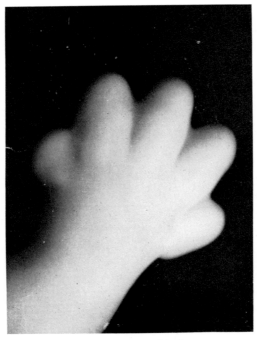

PLATE 10.—¾ webbed

A morphological rating for an embryo is calculated by adding up the numerical values for each of the five developmental features (Table 4). For example, an embryo with forefeet ¼ webbed, hindfeet ¾ webbed, ears ⅓ closed, hair follicles on body and side of head, and eyes open would have a morphological rating of 9.

The morphological conditions themselves are outlined below:

FEET (Plates 9–14). The anterior foot is generally in a slightly more advanced condition than the posterior foot, but both go through the same sequence of stages: (a) unindented plate—a foot that is smooth on its dorsal surface; (b) indented plate—a foot with shallow convolutions on its dorsal surface indicating future digits; (c) plate—a foot with the digits clearly outlined on its dorsal and distal sur-

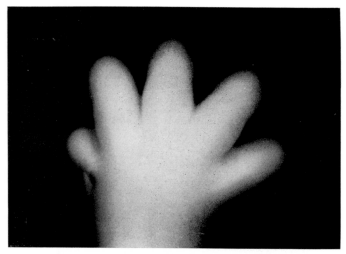

PLATE 11.—½ webbed

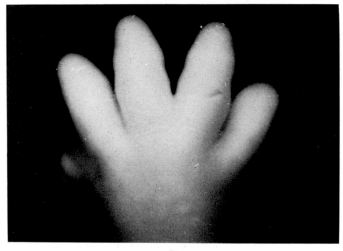

PLATE 12.—Free

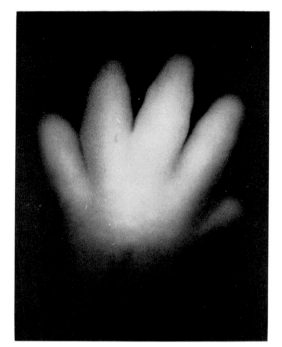

PLATE 13.—¼ fused

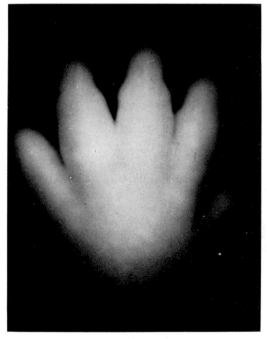

PLATE 14.—½ fused

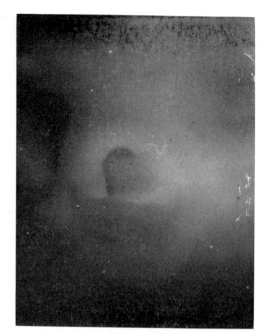

PLATE 15.—Right-angled pinna

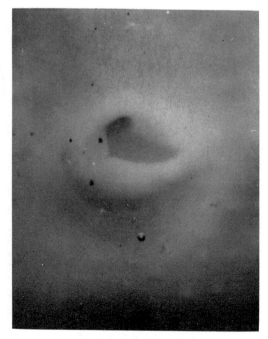

PLATE 16.—⅓ closed

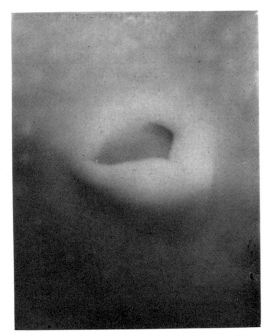

PLATE 17.—½ closed

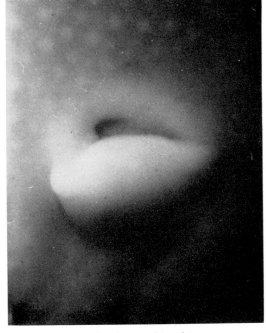

PLATE 18.—⅔ closed

faces; (*d*) ¾ webbed to free—the digits grow more distinct and start to diverge, first the tips of the digits become free, then the separation continues until the digits are completely free from each other; (*e*) ¼ fused to ¾—the digits go from a free divergent condition to where they lie parallel to each other and start to become webbed (henceforth called "fused" to avoid confusion with preceding stage). Fusion starts proximally and extends distally until the digits are ¾ fused. At this point the distal phalanges start to differentiate and no further fusion occurs. "Fused" can be distinguished from "webbed" by long parallel digits as contrasted to stubbier, divergent digits.

EARS (Plates 15–18). The ear pinna arises posterior to the external auditory

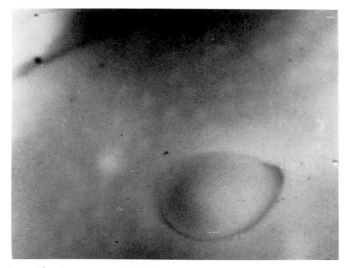

PLATE 19.—Open eye and hair follicles stage "on body and sides of face"

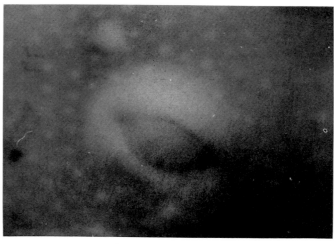

PLATE 20.—½ closed eye

meatus and grows out and then anteriorly until it fuses with the head, thus completely covering the external auditory meatus. Embryos are grouped into the following categories: (*a*) small right angled pinna—a small lump on the side of the head; (*b*) right angled pinna—a small triangle that has not yet started to curl anteriorly; (*c*) ⅓, ½, ⅔ closed—a pinna that has curved forward and has extended across the auditory meatus for approximately the distance indicated; (*d*) ¾ closed—restricted to the condition just preceding fusion, in which a small slit or hole can still be seen between the pinna and the head; (*e*) closed—the condition where the pinna is fused to the head.

HAIR FOLLICLES. The hair follicles first appear on the body, later spreading over the cheeks and finally over the crown of the head. Most of the assigned classes for hair follicles are self-explanatory. "Body and a few on sides of head" refers to the state in which a few follicles occur at the base of the cheeks (Plate 19). "Wrinkled body" is a category that was introduced to distinguish day 17 embryos from day 16 embryos. It is typified by deeper wrinkles and less prominent hair follicles on the body than can be found on the day 16 embryos. For a thorough discussion of how to distinguish day 17 embryos, see Gruneberg (2).

EYES (Plates 19–20). The eyes are wide open through the time of normal palate closure but start to close shortly after. "One-half closed" refers to any stage between open and closed unless the deviations from the latter two conditions are quite minor.

CHAPTER 3

INTERPLAY OF INTRINSIC AND EXTRINSIC FACTORS

HAROLD KALTER

Variability is of concern only to sophisticated cultures in which a certain level of understanding of the grosser aspects of matter has been reached and energy is thus freed that can be devoted to examining the structure and interaction of complex microphenomena.

To the engineer, the quality control expert, the taxonomist, variability means irregularity, something that detracts from the ideal and something therefore to be eliminated. But to the developmental biologist, variability is not something extraneous or even disturbing; rather it is the essence of that which he studies.

Variability in biology has three sources; and, one would imagine, it can be controlled by standardizing and regulating these origins, that is, by creating undeviating and absolute genetic, physiologic, and environmental uniformity. I do not foresee all these objectives being attained so far as the organismic level is concerned, but approximations toward these goals may be achieved.

Let us turn, then, to the first source of variability in abnormal embryonic development—heredity. Dim intimations of the influence of genetic factors in experimental mammalian teratogenesis had been obtained before 1950, in the work of several investigators with rats. But it was not until strains of inbred mice were used in studies of this sort that the full extent of the role of the genotype was appreciated and a rational basis for its investigation was available.

About 15 years ago experiments were started that quickly showed that different strains of mice may respond quite distinctly to the teratogenic action of cortisone (1). This teratogen produces cleft palate, and because of the ease of detecting it and also because of its present-or-absent nature, this defect has lent itself beautifully to studies of the genetics, embryology, and other aspects of experimental teratogenesis.

Research Associate, Children's Hospital Research Foundation, The University of Cincinnati, Cincinnati, Ohio.

Previously unpublished work reported here was partly supported by a grant from the National Association for Retarded Children, New York, New York, and Grant RG-7107 from the National Institute of Arthritis and Metabolic Diseases, The National Institutes of Health, Bethesda, Maryland.

I want to make perfectly clear at the outset that by cleft palate I mean cleft of the palate only, with no involvement of the lip at all. This malformation is seen clearly in Plate 1.

At first, two strains of mice were used, but in the following years a number of others have been treated in the same or similar ways. The highest incidence of cortisone-induced cleft palate, regardless of strain, is produced by four daily injections of 2.5 mg beginning about the eleventh day of pregnancy.

Table 1 shows the incidence of induced cleft palate for five different strains of mice, and we see here a range from a low of 12 per cent for the CBA strain to 100 per cent for the A/J strain (2). Obviously these differences in susceptibility denote genetic differences among these strains, and to investigate the hereditary aspect, the strains were crossed to each other (3).

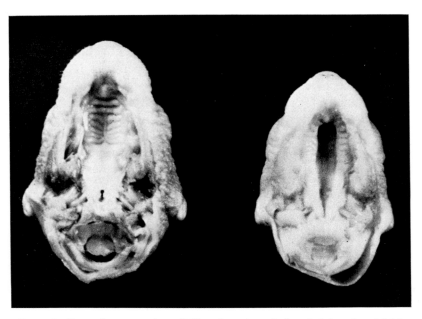

PLATE 1.—Control mouse palate (*left*) and cortisone-induced cleft palate (*right*).

TABLE 1

INCIDENCE OF CORTISONE-IN-
DUCED CLEFT PALATE (CP) IN
STRAINS OF INBRED MICE

Strain	Per Cent CP
CBA	12
C57BL	19
C3H	68
DBA	92
A	100

Harold Kalter   58

Let me stop here and explain briefly something of the genetics of inbreeding. The purpose of inbreeding is to produce a group of organisms all of whom share an identical hereditary composition. The rapidity with which genetic uniformity is accomplished through inbreeding depends on the closeness of relationship between the parents. For mice, the preferred method of inbreeding is mating of full sibs, i.e., brother by sister, which produces nearly 100 per cent genetic uniformity after about twenty generations.

Now we ask, what is the genetic structure of an F_1 or hybrid generation between two inbred strains? The answer is that all such hybrids are gen-

TABLE 2

INCIDENCE OF CLEFT PALATE (CP) IN OFF-
SPRING OF RECIPROCAL CROSSES
BETWEEN INBRED STRAINS

Cross		Per Cent CP
♀	♂	
A	A	100
A	C57BL	44
C57BL	C57BL	19
A	A	100
C3H	A	86
C3H	C3H	68
A	A	100
A	CBA	25
CBA	A	20
CBA	CBA	12
C3H	C3H	68
C3H	CBA	47
CBA	C3H	29
CBA	CBA	12

erally uniform also, and this is so regardless of which of the inbred strains is used as the maternal parent and which as the paternal parent.

Therefore, getting back to our studies, since the offspring of reciprocal crosses between two strains of inbred mice are largely genetically identical, we should expect the incidence of cleft palate in these hybrids also to be identical, if only the genotype of the embryos determines the degree of response to the teratogen. Further it would not be surprising for the hybrid young to have an incidence of cleft palate between the incidences of the pure parental type embryos.

As far as the second point is concerned, Table 2 shows that intermediate frequencies of cleft palate were obtained in many crosses. It also shows, in one case, that offspring of reciprocal crosses were equally susceptible.

But we also see in this table that genetically identical young of reciprocal crosses, though both intermediate in susceptibility to the parent strains, need not have identical frequencies of the malformation. And we note that where this is so, the offspring of the more-susceptible-strain mothers had a significantly higher incidence of the defect than genetically identical offspring of the less-susceptible-strain mothers.

This is further emphasized in Table 3, for here we see that not only were the two types of hybrid young not equal in the incidence of the defect, but they were not even both intermediate. In addition, they showed a yet more striking difference. Let's take a closer look at the data in this table. In the

TABLE 3

INCIDENCE OF CLEFT PALATE (CP) IN OFFSPRING
OF RECIPROCAL CROSSES BETWEEN
INBRED STRAINS

	Cross		Per Cent CP
	♀	♂	
1	A	A	100
2	A	C57BL	44
3	C57BL	A	3
4	C57BL	C57BL	19
5	A	A	100
6	A	C3H	54
7	C3H	A	86
8	C3H	C3H	68
a	A	A	100
	A	C3H	12
	C3H	A	44
	C3H	C3H	36

a Loevy (4).

cross between the A/J and C57BL strains, the offspring of C57BL females by A/J males had a far lower incidence of the defect not only than the young of the reciprocal cross, but also than the pure resistant-strain young.

Two facts emerge from these results. First, not surprisingly, the genetic makeup of the embryo plays a role in the response to the teratogen. This is seen by comparing Cross 1 with 2 and Cross 3 with 4 (Table 3). Here, despite the mothers being of the same strain, the incidences of the defect are quite different, and this must therefore be due to the genetic differences between the offspring. Second, the mother's genetic constitution also plays a role. This is seen in Crosses 2 and 3 where, although the babies' genotypes are the same, differences in susceptibility still exist; and this must thus be the result of these identical embryos having had mothers of different genetic makeup. What emerges, therefore, is that the probability of any embryo's having a cleft palate depends on the hereditary structure of what can be called the maternal-fetal organism acting as a unit.

And we see again from the upper set of crosses (Table 3) that the smallest incidence of cleft palate occurred in hybrid offspring borne by the resistant-strain females. But now let us look at the middle set of figures, from the crosses of the A/J and C3H strains, because here there seems to be a contradiction.

In these results the two hybrid types also differed in frequency of the defect, depending on the strain that was used as the maternal parent. But instead of the lower incidence occurring in the hybrid offspring of the more-resistant-strain mothers, it is the opposite—the lower incidence was in the hybrids of more-susceptible mothers. And to corroborate that this puzzling result was not accidental, the bottom set of figures, from work done in another laboratory (4), is presented. The technique was the same, except the dose was half that given in our work. The incidences of cleft palate were therefore lower, except in the very susceptible A strain, in which it remained 100 per cent. But the point to be noted is that here also the low incidence ocurred in hybrid young from susceptible-strain mothers.

From these strange results we see how difficult it is to generalize about developmental phenomena.

Further genetic analysis of this complex situation was attempted by crossing various parental and hybrid types. The upshot of these experiments can be briefly summarized. Susceptibility to the teratogenic action of cortisone appears to be controlled by several genes; i.e., it is a quantitative characteristic having as its basis genes that may be cumulative in their action, so that the more of them present, all other things being equal, the greater is susceptibility to the teratogen (3).

The difficulties of analyzing a quantitative hereditary trait stem greatly from the fact that such traits are easily influenced by non-genetic factors. Studies of non-genetic modification of the teratogenic effects of cortisone have but barely begun to reveal the net of interacting forces at work. The next source of variability, then, is environment.

There is time here only to touch on the few environmental sources of variability that have been investigated. First, certain maternal attributes were considered, e.g., parity, age, and weight (5–7). Treated females were allowed to deliver and were remated several times (5); as can be seen in Table 4, the incidence of cleft palate steadily decreased with advancing parity. However, since parity is closely correlated with maternal age, it was necessary to determine the possible effect of this quality. To obviate the parity effect, all females (6, 7) were primiparous. Table 5 shows that, indeed, even though all the females were of uniform parity, there was a fairly good relationship between maternal age and incidence of cleft palate, the one increasing as the other decreased, in two groups of different genetic composition. But age in mice is further associated with weight, and, as Table 6 shows, this parameter was also closely inversely associated with frequency of the defect.

And now we ask, are maternal age and weight both really associated with incidence of the defect, or does one or the other only appear to be associated because they are closely correlated with each other?

TABLE 4

PER CENT OF CLEFT PALATE ACCORDING
TO LITTER SERIATION

PARITY				TOTAL
1–2	3–4	5–6	7–9	
26	19	13	5	20

TABLE 5

INCIDENCE OF CLEFT PALATE (CP) IN OFF-
SPRING OF PRIMIPAROUS MICE ACCORDING
TO MATERNAL AGE AT CONCEPTION

GROUP 1		GROUP 2	
Mean Maternal Age (days)	Per Cent CP	Mean Maternal Age (days)	Per Cent CP
78	70	106	96
99	56	200	74
117	60	242	79
148	36	405	92

TABLE 6

INCIDENCE OF CLEFT PALATE (CP) IN OFF-
SPRING OF PRIMIPAROUS MICE ACCORD-
ING TO MATERNAL WEIGHT AT CONCEP-
TION

GROUP 1		GROUP 2	
Mean Maternal Wt. (g)	Per Cent CP	Mean Maternal Wt. (g)	Per Cent CP
21	78	22	94
23	60	27	81
25	47	30	79
28	39	33	78

This question is readily answered by the use of the multiple regression technique, which holds all variables constant but one and tests for association with that one. By this means it was found, in three separate studies using mice of three different genotypes, that maternal weight and not maternal age was significantly associated with frequency of cleft palate (6–8).

One may comment that of course there is an inverse relation between maternal weight and incidence of the abnormality, since a uniform dose of cortisone was given in these studies. And indeed, in recent studies, in which dosage of steroid was based on weight, the influence of weight was largely eliminated (9).

But the lesson we can learn from this association of maternal weight and incidence of cleft palate is not weakened by its being largely due to a uniform dose of teratogen having been given.

First, we have learned that the fetus is responsive to a maternal-dose relationship, thus underscoring the idea that the fetus, in this study, is merely one compartment of a dual organism.

And second, it points to the possible parallel situation in women, and leads one to wonder whether, if as much attention were paid to the weight of women as to their age, some significant relationships might not be uncovered with frequency of some congenital malformations. It should be remembered, of course, that the weight increase that goes with aging in mice is mostly due to fat accumulation. Investigations of similar phenomena in women must not lose sight of this fact and therefore should take into consideration not merely body weight but body composition as well.

Not only maternal weight but fetal weight as well is closely associated with cleft palate. The steep inverse association between these can be seen in Figure 1. These results are from an experiment in which a uniform dose of cortisone was given (2).

But although the relationship between maternal weight and cleft palate was largely eliminated by experiments in which steroid was given on a weight basis, no such disappearance occurred so far as fetal weight was concerned, since this continued to be significantly inversely related to frequency of cleft palate. Figure 2 shows this fact for two groups that received a smaller and a larger dose of hormone per gram of maternal weight (9).

What is to be understood by this new association? First of all, if there is a causal relationship here, which is cause and which effect? Are lighter embryos more susceptible or are defective offspring more retarded? We cannot say for sure, of course, but it seems likely that small developmental variations present at crucial embryonic stages can have far greater ramifications than would be expected, whereas large disturbances at less critical moments may have unimportant effects.

This new association brings up all the things by which fetal weight can be influenced. Maternal weight is one of these, but as was said even when the effect of maternal weight was largely abolished, in studies where the dose was given on a weight basis, the relationship of fetal weight and cleft palate was maintained. But numerous other variables are known or suspected of influencing fetal weight, such as litter size, maternal nutrition, fetal uterine position, and so on; and these may, therefore, also affect incidence of cleft palate.

The evidence for the influence of litter size on cleft palate is contradictory. In an earlier study (6) no such effect was found when maternal weight was

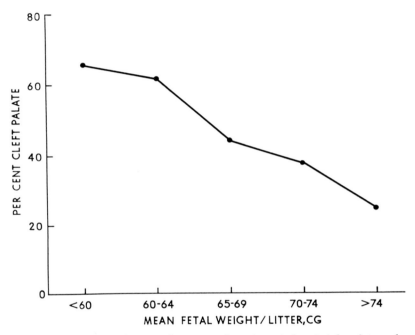

FIG. 1.—Inverse relationship between frequency of cortisone-induced cleft palate and mean fetal weight. From observations of a study in which a uniform dose of hormone was given.

held constant. But in more recent studies (9) in which, instead of litter size, the number of young per uterine horn was considered, an effect was noted. Figure 3 shows that as the number of live inhabitants of a horn increased, the frequency of cleft palate also increased. This correlation may be understood when it is remembered that increased litter size is accompanied by reduced average weight.

However, one other crucial observation is not in harmony with this finding. As others have noted, it was found that the number of offspring in the two uterine horns are inversely correlated (Fig. 4). We are faced, then, with two facts: (1) within a uterine horn, the fetal number and cleft palate frequency were positively correlated, and (2) between uterine horns the fetal numbers were inversely correlated. Putting these facts together, one should also expect an inverse relationship between the horns in frequency of cleft palate. A relationship was found; but, as Figure 5 shows, instead of a negative relationship, it was a positive one. This, too, cannot be explained yet.

Before we are led too far astray by pursuing this tangled trail, let us turn to a very different environmental source of variability—season. The data for one specific type of cross were collected intermittently for 6 years; and when plotted by month of the year showed, as seen in Figure 6, that, generally, a higher incidence of cleft palate occurred during winter months than summer ones (10). The seasonal variation can be considered to take the form of a sine curve, such as was drawn free-hand here.

One way of handling data such as these is to compare periods with each

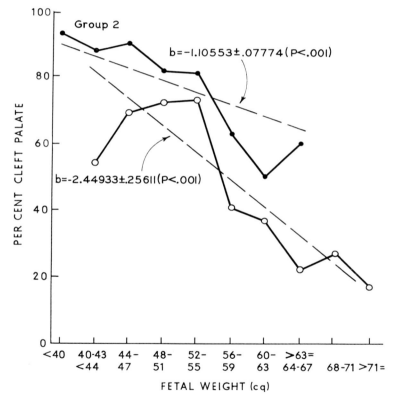

Fig. 2.—Here, as in the study whose results are presented in Figure 1, an inverse relationship between fetal weight and incidence of cleft palate was noted. However, the data recorded in this figure show that this relationship occurred despite the teratogen being given on a maternal weight basis: a daily injection of 0.10 mg (Group 1) or 0.0625 mg (Group 2) prednisolone acetate per gram of weight at conception for 4 days beginning 11 days after observing the vaginal plug. The regression coefficients indicate a 1.10 per cent and 2.45 per cent decrease in frequency of the abnormality with each 0.01 g increase in fetal weight.

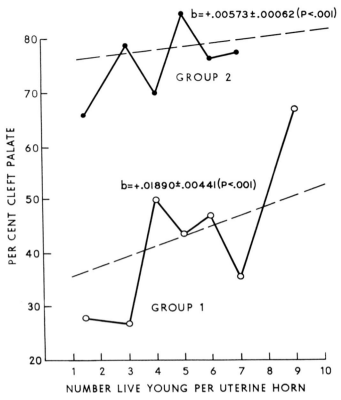

FIG. 3.—The regression equations indicate that in the study described in Figure 2 there occurred statistically significant positive associations between the number of live young per uterine horn and the incidence of cleft palate in the horn.

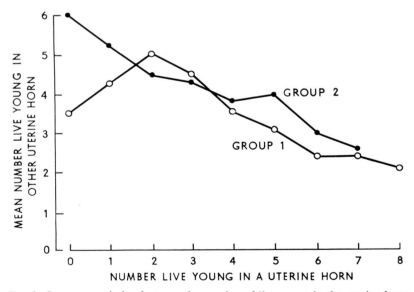

FIG. 4.—Inverse association between the number of live young in the uterine horns.

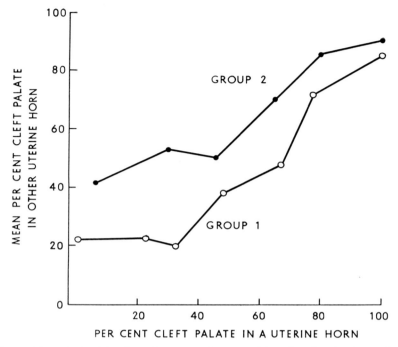

Fig. 5.—Positive relationship of the frequency of young with cleft palate in the uterine horns.

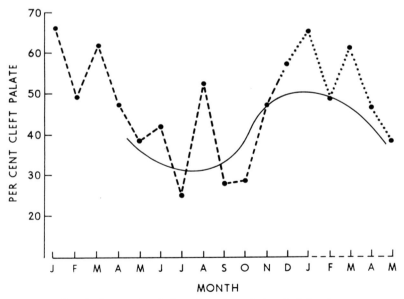

Fig. 6.—Seasonal variation of cortisone-induced cleft palate.

other. In this case the 6-month winter-centered period of November to April was compared with the warm months May to October. This comparison (Table 7) demonstrates that a significantly higher incidence of the defect occurred in the winter months than the summer ones. Also notice that no part of this difference was due to seasonal variations in maternal weight (10).

We can only guess what the causes of such a seasonal variation could be. Temperature is of course the first to be suspected; but would not one expect warm summer temperatures to be catalytic and produce high, not low, frequencies? Humidity also varies seasonally, being high in the summer and low during the winter months. In the past several years we have controlled this fluctuation by humidifying the air in the winter and dehumidifying it in the summer. Another possible seasonable variable is composition of the diet. We are assured by the manufacturers that the diets we use do not vary, but this

TABLE 7

INCIDENCE OF CLEFT PALATE (CP)
ACCORDING TO MONTH

	Nov.–Apr.	May–Oct.	Total
No. females.........	163	143	306
Total young.........	1,147	1,109	2,256
No. with CP........	612	400	1,012
Per cent CP........	53.3	36.1	44.9
Mean maternal wt...	23.0	22.8	22.9

aspect still deserves scrutiny. The question of the possible influence of diet is an important one (see chap. 2), and so I will discuss some studies done with this factor.

First, of course, it was necessary to learn what the administration of cortisone may do to the amount of food eaten by pregnant mice. Might some of the assumed effects of cortisone be due, actually, to suppression of appetite? It was found not to be (11), since cortisone, in fact, increased the appetite; so that, even though causing a smaller increase in weight during pregnancy, the amount of food eaten was increased (Fig. 7).

If cortisone increases food consumption, what would be the effect of depriving pregnant animals of food? And, further, what would be the effect of a combination of a restricted diet plus a small dose of cortisone, a dose that by itself did little?

An experiment to supply some answers was set up (12), and five groups of animals were treated as follows: one group received 0.5 mg of cortisone; the second, 1.0 mg; the third, 2.5 g of food per day for 5 days; this is about 40 per cent of the normal average intake; the fourth and fifth groups received one dose or other of cortisone plus the restricted caloric intake (Table 8).

Figure 8 shows what effect these treatments had on maternal weight gain and the incidence of cleft palate produced by them. Note that, most surprising of all, the restricted diet alone produced 5.6 per cent cleft palate and that a great potentiation was produced by the combined treatments. Here then is evidence that a significant role in the teratogenic action of cortisone is played by quantity of diet. But what about quality?

To investigate this aspect mice were fed (13) the regular laboratory diet supplemented by large amounts of riboflavin, folic acid, protein, or dextrose. In no case did a supplemented group have an incidence of cleft palate significantly different from its control (Table 9). These ingredients, therefore, or the manner of administering them, were ineffective in modifying the action of cortisone.

TABLE 8

Group	Treatment
1	0.5 mg cortisone, ×4
2	1.0 mg cortisone, ×4
3	2.5 g food, ×5
4	0.5 mg cortisone, ×4, plus 2.5 g food, ×5
5	1.0 mg cortisone, ×4, plus 2.5 g food, ×5

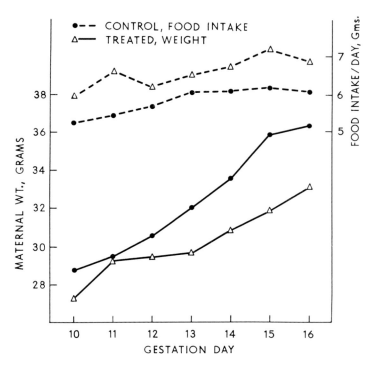

FIG. 7.—Effect of cortisone on food consumption and weight gain of pregnant mice.

Intrinsic and Extrinsic Factors

But to consider this experiment of negative results as a wasted effort would be to overlook its valuable lesson. Notice that no experimental group differed significantly from its control; but also notice how variable the control values were, with the lowest one significantly different from the highest.

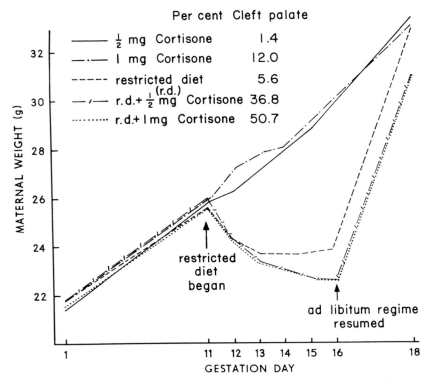

Fig. 8.—Effect of hypocaloric diet and/or small doses of cortisone on weight gain of pregnant mice and on incidence of cleft palate.

TABLE 9

EFFECT OF SUPPLEMENTATION ON INCIDENCE
OF CLEFT PALATE (CP)

Group	Treatment	Females	Offspring	Per Cent CP
1	Cortisone......................	18	127	31
	Cortisone+riboflavin (1 g/kg).....	25	199	34
	Cortisone+riboflavin (250 mg/kg)	70	547	31
2	Cortisone......................	24	181	42
	Cortisone+folic acid (200 mg/kg)..	28	209	42
3	Cortisone......................	16	116	54
	Cortisone+dextrose.............	19	131	65
4	Cortisone......................	17	107	58
	Cortisone+protein.............	18	112	56

Think of the trap that would have been fallen into if Control 4 had been considered good enough and a separate control not obtained for Group 1. It would have appeared that the riboflavin supplementation caused a significant reduction in the incidence of cleft palate.

This is exactly the sort of thing of which one must be aware. To produce congenital malformations in experimental animals, as Dr. Wilson mentioned in his introductory remarks, has proven very easy. After the way was shown us by the great pioneering efforts, it has not required much skill or care to demonstrate that malformations are inducible in mammals and that they are attributable to the maternal insult. But an important part of teratological experimentation in the future will be of a different sort. It will consist of the study of combined treatments, for the purpose of throwing light upon the mechanism of a teratogen through the prevention or exacerbation of its effects. And if we do not sufficiently respect the intrinsic lability of developmental processes, we are going to be deceived time and again into thinking that something we have done is always responsible for the differences that may occur among groups of differently treated animals. To avoid these traps we are going to need to be far more concerned than we have been with the genetic and environmental definability and uniformity of the materials of our experiments.

Let me give you some examples. Between 1956 and 1960, Millen and Woollam, working in Cambridge, England, published a series of papers that dealt with the results of various combined teratogenic treatments (14). Working with rats, these investigators found, as many did before and have since, that cortisone is not teratogenic in that species. Yet when cortisone was administered to pregnant rats that were also given large amounts of vitamin A, the combined treatment appeared to intensify the effects of the vitamin given alone.

Treatment of pregnant rats with excess vitamin A plus several other substances also often resulted in modification of the effects of the vitamin by itself, the frequency of malformations being sometimes augmented, sometimes diminished.

Although these results were greeted skeptically, nothing was done about it until Cohlan in New York City attempted in a large-scale experiment to confirm some of these findings. Cohlan found that in no instance of a combined treatment was the result different from that obtained with excess vitamin alone (15).

What were the reasons for these discrepancies? Of course, no one can say for sure; but there is every reason to believe that in addition to the major variables we are well aware of, such as strain, stock, and species differences, and differences in environmental conditions and in methodology, there are

Intrinsic and Extrinsic Factors

other variables, which are inconspicuous and transient, and of which we may be totally unaware.

To minimize these influences our studies cannot afford to be short-term and makeshift. We must take every possible precaution to insure that at least the most obvious variables are randomized, and that as little as possible has been overlooked that may cast doubt on the validity of our results.

I do not think it can be stressed too much that credible comparative studies of combined teratogenic factors require the utmost skill, patience, and effort.

Although the importance of the macroenvironment cannot be doubted, it is to the microenvironment that we look for some real secrets. This term can mean not only such temporal factors as sequential patterns of biochemical and embryological events, but also spatial factors, though they may be as prosaic as the relative position of fetuses in the uterine horn with respect to

TABLE 10

CLEFT PALATE FREQUENCY OF FETUSES LYING NEXT TO
THOSE WITH AND WITHOUT CLEFT PALATE

LOCATION	GROUP 1		GROUP 2	
	No. with CP	Per Cent CP	No. with CP	Per Cent CP
Next to CP..........	133	58.3	357	85.4
Next to non-CP......	95	32.4	63	50.0

some landmark such as the ovary or to each other. I will give a couple of examples of what appear to be the working of such a factor.

In Table 10 is shown the frequency of cleft palate according to whether the young lay next to littermates with or without the defect. Those located next to abnormal ones have a significantly higher incidence themselves than those located next to normal young (9).

That the relative position of fetuses may be subtle enough to influence their weight is seen in Table 11. Here are compared the weights of normal young according to whether located next to an affected or a normal littermate; and it is seen that those next to normal ones are significantly heavier than those next to malformed babies. But even more surprising, the farther away a normal young is from an abnormal one, the heavier it is (Table 12).

This is as far as I can go at this time with this subject; and now I would like to turn to another teratogen whose effects are very different from those of cortisone. These effects are far more complex in many ways and hence their genetics and other factors influencing them have proven even more difficult to investigate than cortisone-induced cleft palate.

Acute, temporary deficiency of riboflavin, induced by the antagonist, galactoflavin, was first shown by Nelson and co-workers in 1956 to be teratogenic in rats (16). Its teratogenic effects in mice are similar to those in rats in producing extensive defects of the skeleton and several soft-tissue anomalies, but there are important species differences as well. I will concentrate on the aspects of this work that are best investigated by the use of inbred animals.

Four strains of mice were treated with two dose levels of the riboflavin

TABLE 11

MEAN WEIGHT OF NORMAL FETUSES LYING NEXT TO THOSE WITH
AND THOSE WITHOUT CLEFT PALATE (CP)

| Group | Normal Fetuses Lying Next to | | | |
| | CP | | Non-CP | |
	No.	wt (g)	No.	wt (g)
1	90	0.5983 ± .0066	168	0.6337 ± .0051
2	55	0.5373 ± .0084	41	0.5656 ± .0092

TABLE 12

MEAN WEIGHT OF NORMAL FETUSES
LOCATED ONE TO FOUR IMPLANTA-
TIONS FROM THOSE WITH CP, WITH
NO RESORBED SITES INTERVENING

Implantation Distance from CP Fetuses	No.	Weight (g)
1.........	31	0.5871
2.........	13	.5997
3.........	11	.6344
4.........	6	0.6433

antagonist. The four strains were the A/J, DBA, 129, and C57BL. They were fed a mixture of a commercially obtained riboflavin-deficient diet and either 60 or 90 mg/kg of galactoflavin. These diets were given for 4 days beginning early on the 10th or 11th day of pregnancy (17).

The results were very complex in several respects. First, some malformations were variable in severity, site, or laterality. These points can be illustrated by one of the malformations, syndactyly of the toes. As Table 13 shows, this defect consisted of variable degrees of fusion of certain digits, in most cases of the third and fourth toes. Most abnormal offspring were bilaterally affected, but in unilateral cases, the left foot was more often abnormal than the right. This table also anticipates the strain and other differences in incidence of malformations that will be discussed.

In addition to such variability, several forms of strain differences made the results complex. First, the strains differed in response to the two dose levels of antagonist. From an over-all point of view, the DBA and 129 strains were highly susceptible to both dose levels, the A/J strain was only slightly susceptible to the smaller dose but very susceptible to the larger, and the C57BL strain was not at all susceptible to the smaller and only slightly susceptible to the larger.

Next there were qualitative strain differences. That is, some strains exhibited certain malformations not found in the three other strains. And last, the strains differed widely in the incidence of individual malformations and defects, so that, in sum, each strain showed its own frequency, severity, and spectrum of congenital malformations.

In the next three tables are examples of these quantitative strain differences. First (Table 14) are limb defects; and next (Table 15) those of the face and mouth. Skeletal defects of almost every bone were found. Internal

TABLE 13

GALACTOFLAVIN-INDUCED SYNDACTYLY OF TOES, NUMBER AND PER CENT

Cross	Total Abn.	Left		Per Cent Left	Right			Per Cent Right	Per Cent Uni-lateral	Bilateral				Per Cent Bilateral
		2–3	3–4		2–3	3–4	2–4			2–3	3–4	2–4	Asymm.	
A×A	153	1	4	3	0	1	0	1	4	0	132	12	3	96
B×B	48	1	5	12	0	1	3	8	20	0	5	18	15	80
D×D	99	1	17	18	2	4	0	6	24	1	62	11	1	76
A×B	81	0	14	17	0	3	0	4	21	0	58	1	5	79
B×A	25	0	6	24	0	0	0	0	24	0	18	0	1	76
A×D	80	0	9	11	0	4	0	5	16	0	66	1	0	84
D×A	69	0	11	16	2	0	2	6	22	2	47	4	1	78
B×D	19	0	1	5	0	1	0	5	10	1	7	7	2	90
D×B	40	3	7	25	2	2	0	10	35	6	17	2	1	65

TABLE 14

STRAIN DIFFERENCES IN FREQUENCY OF GALACTO-
FLAVIN-INDUCED LIMB DEFECTS

Strain	Micromelia of Arm (%)	Ulnar Abduction of Hand (%)	Syn. and/or Oligo-dactyly (%)	Misc.[a] (%)
A	19	17	27	7
DBA	39	31	5	18
129	18	24	82	8
C57BL	13	12	8	5

[a] Short digits, abnormal paws, malposition of feet, etc.

malformations consisted mostly of abnormalities of the brain and absence of the esophagus (Table 16).

These strain differences, as do those of cortisone-induced cleft palate, indicate that the response to the teratogen is at least partly under the control of genetic influences. The bases of these influences were investigated, as in the cortisone studies, by crossing the strains. For this purpose only the larger dose of galactoflavin was given and three strains used, the A/J, C57BL, and DBA, and they and their hybrids were crossed in all possible relevant combinations (18, 19).

TABLE 15

STRAIN DIFFERENCES IN FREQUENCY OF GALACTO-
FLAVIN-INDUCED FACIAL DEFECTS

Strain	Open Eye (%)	Cleft Palate (%)	Brachy- gnathia (%)
A	6	3	1
DBA	4	41	40
129	2	8	1
C57BL	18	13	10

TABLE 16

STRAIN DIFFERENCES IN FREQUENCY OF
GALACTOFLAVIN-INDUCED INTERNAL
MALFORMATIONS

Strain	Brain Abnormalities (%)	Absent Esophagus (%)
A	55	78
DBA	44	83
129	9	30
C57BL	5	2

The first column of Table 17 shows the types of crosses made. The first three crosses are within-strain crosses and the next six, reciprocal between-strain crosses. Then come six that are called backcrosses since they are of first generation hybrids crossed to parental types. And the bottom three lines are crosses of first generation hybrids to each other.

The body of the table lists only the most frequent abnormalities for which the young were scored. In almost every reciprocal cross, hybrids with more-susceptible-type mothers were more frequently affected with specific anomalies than genetically identical ones with less susceptible mothers. These maternal influences immediately alert us to the fact that, once again, as in

75   *Intrinsic and Extrinsic Factors*

the cortisone studies, the mother's genetic constitution plays a role in modifying the fetal response to galactoflavin. It is only fair to point out that fathers play a role also. The paternal contribution is conveniently seen in the backcrosses, in which each type of genetically identical hybrid mother was bred to two types of males. And we can easily see in these crosses that the more frequently defective young were usually those with the more-susceptible-strain fathers.

Inspection of this table also reveals that certain patterns of susceptibility

TABLE 17

Per Cent Offspring with Specific Galactoflavin-induced Congenital Defects

Cross (♀×♂)	No. Litters	Total Offspring	Synd. Toes	Synd. Fingers	Bent Hand	Club-foot	Short Arm	Open Eye	Cleft Palate	Brachy-gnathia	Abn. Brain	Absent Esoph.
A×A	22	191	80	20	40	2	14	22	3	1	80	54
B×B	21	163	29	0	5	2	0	6	0	1	5	2
D×D	23	178	56	4	75	41	56	17	61	36	84	78
A×B	21	160	51	2	6	2	6	1	1	1	19	12
B×A	21	120	21	0	0	1	0	0	1	0	2	5
A×D	26	198	40	2	2	1	1	2	1	0	8	28
D×A	22	172	40	8	37	8	16	10	12	6	55	57
B×D	20	115	16	4	6	6	4	8	5	3	16	10
D×B	23	89	45	3	49	16	34	18	36	14	21	73
AB×A	22	158	32	1	3	1	0	1	0	0	22	23
AB×B	23	175	21	1	5	0	0	0	0	0	11	7
AD×A	20	126	48	1	11	2	3	2	3	5	30	47
AD×D	20	148	10	1	16	3	4	2	11	14	29	37
BD×B	23	204	1	0	0	1	0	2	0	2	7	2
BD×D	23	196	1	0	3	6	0	3	12	14	24	12
AB×AB	30	219	23	0	2	0	0	0	0	0	13	4
AD×AD	19	147	8	1	5	1	2	1	5	10	52	33
BD×BD	26	209	3	1	4	0	0	2	0	0	22	2
	405	2,968										

existed, but that a cross with a high incidence of some defects did not necessarily have a high incidence of all defects.

We see then that the genetic qualities of three individuals are involved here: mother, father, and fetus. And, as was noted above, the fetus is more than just a combination of the genetic contributions of its parents, since its role does not concern itself by itself only, but also as part of a dual unit. How, then, can a genetic analysis of this complex situation be made, especially when the signs of its existence—congenital malformations—are numerous and are both quantitatively and qualitatively variable?

The solution may lie in simplifying the situation by considering the abnormalities as comprising a number of syndromes. A syndrome can be considered here as a specific combination of anomalies that appear together be-

Harold Kalter   76

cause they have a common chemico-embryological cause and are thus the expression of a single genetically determined response.

The simplest and most conveniently handled assumption is that all the defects, regardless of their incongruity and variety, are parts of just one syndrome. This may be an oversimplification, since, e.g., even susceptible mice have extremely low incidences of some defects. But ignoring this fact, what can be done with the data if all the abnormalities are lumped?

Table 18 shows some pertinent figures of this sort, such as per cent abnor-

TABLE 18

FREQUENCY OF GALACTOFLAVIN-INDUCED CONGENITAL DEFECTS

Cross (♀ × ♂)	Total Young	Abnormal Young (%)	Mean No. Defects/ Abn. Young	Per Cent Abnormal Young with		
				1 Defect	2 Defects	> 2 Defects
A×A	191	90	3.8	11	19	70
B×B	163	38	1.5	69	18	13
D×D	178	90	6.1	10	6	84
A×B	160	61	1.7	73	12	15
B×A	120	24	1.3	69	31	0
A×D	198	48	1.9	42	43	15
D×A	172	72	3.8	26	14	60
B×D	115	31	2.8	39	19	42
D×B	89	81	4.4	17	11	72
AB×A	158	46	1.8	44	38	18
AB×B	175	30	1.5	74	9	17
AD×A	126	72	2.2	38	32	30
AD×D	148	52	2.5	38	26	36
BD×B	204	12	1.6	58	25	17
BD×D	196	35	2.2	38	28	34
AB×AB	219	30	1.4	69	23	8
AD×AD	147	56	2.1	42	30	28
BD×BD	209	24	1.6	68	16	16

mal offspring, mean number of defects per abnormal offspring, and per cent of abnormal young with different numbers of defects. We see that for such data there are also strain differences. Thus the C57BL strain has a much lower incidence of abnormal young than the other two strains. It appears, however, as though the A/J and DBA mice have equal percentages of abnormal offspring, but if we look at the next column it is seen that abnormal DBA's have a far larger mean number of defects than A/J young. This, therefore, is a further index of strain difference. And looking to the right side of the table we find a definite relationship between degree of over-all susceptibility and frequency of multiply malformed offspring.

Going down to the reciprocal strain crosses, we see these indices of susceptibility are also subject to maternal influences. It therefore appears that there is a parallelism between degree of susceptibility as expressed by incidence of particular malformations and by over-all incidence of abnormality, and hence that the latter is a legitimate statistic to use in further analyses of the problem.

No intense analyses of these data have yet been made, but a beginning has been made, however, and the data are being tested for certain possible associations.

First, what about the relationship between fetal death and fetal abnormality? It has often been assumed in teratological studies that fetal death is the extreme action of teratogens but not qualitatively a different expression of that action. If this were so, the lethal effect of prenatal disturbances would increase with its teratogenic potency.

Although this may be true for some teratogens, it did not hold in this study

TABLE 19

EFFECTS OF GALACTOFLAVIN
VS. CORTISONE

| STRAIN | PER CENT CP PRODUCED BY | |
	Galactoflavin	Cortisone
A/J	3	100
C57BL	0	19
DBA	61	92

since no correlation at all was found between the fetal death rate and the incidence of malformed offspring among the eighteen different crosses made. It appears, therefore, that galactoflavin-induced death and galactoflavin-induced malformations are unrelated effects.

And what about maternal weight and abnormality? In the present study, maternal weight was not associated with the over-all incidence of abnormality, nor was it related to the frequency of certain particular malformations with which it was tested, such as cleft palate, syndactyly, absent esophagus, etc.

One more thing deserves attention, and that is the contrast between the action of cortisone and galactoflavin on the palate. Table 19 shows the incidence of cleft palate induced by these two teratogens in three strains. The table succinctly discloses that hereditary susceptibility to one teratogen does not necessarily extend to another and makes us aware of the fact that each teratogenic situation is probably under the control of a distinct genetic mechanism.

These studies with inbred strains of mice have uncovered numerous things that could not have been so clearly demonstrated with other types of experimental material:

First, the undeniable fact that hereditary potential largely determines the quantitative and qualitative ramifications of given prenatal disturbances.

Second, that the nature of this hereditary potential is exceedingly involved, since it is polygenic and involves the fetus and the maternal organism, individually and combined.

Third, that the embryological flow of events, being delicately balanced, is easily disturbed by certain mixtures of circumstances, and yet can sometimes be homeostatically restored to its proper channels.

And, last, that there exists a complex interplay among genes and environment, making subtle, inconspicuous, shifting factors of paramount importance in the development of unborn creatures.

References

1. Fraser, F. C. and Fainstat, T. D. Production of congenital defects in the offspring of pregnant mice treated with cortisone. *Pediatrics* **8**: 527, 1951.
2. Kalter, H. Unpublished observations.
3. Kalter, H. The inheritance of susceptibility to the teratogenic action of cortisone in mice. *Genetics* **39**: 185, 1954.
4. Loevy, H. Genetic influences on induced cleft palate in different strains of mice. *Anat. Rec.* **145**: 117, 1963.
5. Kalter, H. and Fraser, F. C. The modification of the teratogenic action of cortisone by parity. *Science* **118**: 625, 1953.
6. Kalter, H. Modification of the teratogenic action of cortisone in mice by maternal age, maternal weight, and litter size. *Am. J. Physiol.* **185**: 65, 1956.
7. Kalter, H. Factors influencing the frequency of cortisone-induced cleft palate in mice. *J. Exp. Zool.* **134**: 449, 1957.
8. Kalter, H. Further evidence of the association between maternal weight and frequency of cleft palate in the offspring of cortisone-treated pregnant female mice. *Genetics* **42**: 380 (abst.), 1957.
9. Kalter, H. Paper presented at the 3d Annual Meeting of the Teratology Society, Ste Adele, Quebec, 1963.
10. Kalter, H. Seasonal variation in frequency of cortisone-induced cleft palate in mice. *Genetics* **44**: 78 (abst.), 1959.
11. Kalter, H. The effect of cortisone on the food consumption of pregnant mice. *Canad. J. Biochem. Physiol.* **33**:767, 1955.
12. Kalter, H. Teratogenic action of a hypocaloric diet and small doses of cortisone. *Proc. Soc. Exp. Biol. Med.* **104**: 518, 1960.
13. Kalter, H. Attempts to modify the frequency of cortisone-induced cleft palate in mice by vitamin, carbohydrate, and protein supplementation. *Plastic Reconstr. Surg.* **24**: 498, 1959.
14. Woollam, D. H. M. and Millen, J. W. The modification of the activity of certain agents exerting a deleterious effect on the development of the mammalian embryo. In: *Ciba foundation symposium on congenital malformations,* edited by G. E. W. Wolstenholme and C. M. O'Connor. Boston: Little, Brown and Co., 1960, pp. 158–72.
15. Cohlan, S. Q. and Stone, S. M. Observations on the effect of experimental endocrine procedures on the teratogenic action of hypervitaminosis A in the rat. *Biol Neonat.* **3**: 330, 1961.

16. NELSON, M. M., BAIRD, C. D. C., WRIGHT, H. V., AND EVANS, H. M. Multiple congenital abnormalities in the rat resulting from riboflavin deficiency induced by the antimetabolite galactoflavin. *J. Nutr.* **58**: 125, 1956.

17. KALTER, H. AND WARKANY, J. Congenital malformations in inbred strains of mice induced by riboflavin-deficient, galactoflavin-containing diets. *J. Exp. Zool.* **136**: 531, 1957.

18. KALTER, H. Studies on the inheritance of susceptibility to congenital malformations produced by riboflavin deficiency in mice. *Proc. Tenth Intern. Cong. Genet.* **2**: 139 (abst.), 1958.

19. KALTER, H. Unpublished observations.

REPRODUCTION AND GROWTH OF MICE AND MAINTENANCE OF AN INBRED MOUSE COLONY

THE ESTROUS CYCLE

Ovulation occurs, independently of copulation, at fairly regular intervals, the cycles usually ranging from 4–6 days. The ovarian cycles are correlated with the usual changes in the vagina, uterus, and Fallopian tubes, but the external signs are unreliable. The cycle is divided into several periods: proestrus; estrus, during which "heat" occurs; early and late metestrus; and diestrus, the period of rest. The rest period lasts about 2 days, each of the others about a day, but there may be great variation. In old females the cycles tend to become lengthened and irregular.

Mating takes place only during the time of heat, which lasts about 12 hours. Willingness to mate is associated with ovulation, and ovulation and other estrous phenomena are closely linked with the diurnal light cycle. This has been demonstrated by artificially reversing times of light and darkness, which produces a corresponding switch in the estrous cycle. It has been found that ovulation in a group of mice begins about 6 hours after the lights are turned off or the sun sets and is completed in the whole group in about another 6 hours. The time taken by any one mouse to complete ovulation is about 1 hour, and the average interval between ovulation and fertilization is about 5 hours. The influence of light is mediated through the eyes and the central nervous system, which activates the pituitary to secrete the ovulating hormone.

THE VAGINAL PLUG

After copulation, a secretion of the seminal vesicles in the ejaculate of the male coagulates to form a plug in the vagina extending from the cervix to the vulva, where it is ordinarily visible, and is therefore a convenient sign that mating has occurred. Sometimes the plug does not reach the surface and it can be missed, but gentle introduction of a probe into the vagina usually will disclose its presence in these cases. The plug is gradually loosened by leukocytic action and is extruded after about a day.

DURATION OF PREGNANCY

The gestation period lasts about 19 days—occasionally 20 days. In nursing females this is prolonged for 1–2 weeks, due to delay in uterine implantation of blastocysts. During this delay, the blastulas lie free in the lumen of the uterus. The young are most frequently born in the early hours of the morning, between midnight and 4 A.M.

PRENATAL GROWTH

Fertilization occurs in the Fallopian tubes. The cleaving eggs enter the uterus about 3–4 days after mating and become implanted in the uterine

TABLE 1

PRENATAL GROWTH OF MICE

Gestation (day)	Average Weight (mg)
8	0.1
9	1.5
10	9
11	33
12	76
13	130
14	229
15	365
16	593
17	847
18	1,190

mucosa at about the end of the fifth and beginning of the sixth day. At that time the blastula stage has been reached. Hardly any increase in size occurs before implantation, but afterward there is a great explosion in growth, as Table 1 shows. The crown-rump length of an embryo is not a satisfactory index of its age, because of variability and overlap for different ages. A much better system of aging is based on the development of certain external features, such as eyes, ears, toes, and tactile hair follicles. Plate 1 shows prenatal mice at 24-hr intervals beginning with one of the 10th gestation day. Litter size and birth weight vary greatly, but, in general, litters of inbred mice generally have 6–8 young, and newborn weight is approximately 1.2 to 1.4 gm.

EXTERNAL FEATURES OF THE NEWBORN

Mice are born with the eyelids fused together. The external ears are bent forward and firmly attached to the skin of the face and cheeks. Except for vibrissae, they are naked, and the only sign of pigmentation is a ring in the iris, which is clearly visible through the closed eyelids. Of course, even this pigment is absent in albinos.

At birth the sexes can be distinguished quite easily. In both sexes there is a small prominence, called the genital papilla, anterior to the anus. This is somewhat larger in males than females; but much more reliable is the distance between the papilla and anus, which is larger in males. However, this distance varies from litter to litter, and so the safer method is to compare animals within a litter (Plates 2–4).

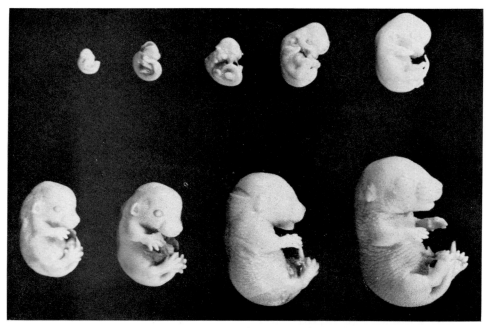

PLATE 1.—Prenatal stages in mice at 24-hr intervals, beginning with 10-day embryo.

POSTNATAL GROWTH

Hairs make their appearance within the skin on about the third day, and if they are pigmented, the skin is darkened. The ears become detached at 4–6 days and are short and fleshy (Plates 5–7). The growth of the baby fur is complete by about 8–10 days (Plate 8). At this time the sexes can again be told apart easily by the presence of five pairs of conspicuous nipples in the female; but these tend to become obscured in a few days by the lengthening hairs.

At about 11 to 13 days the incisors erupt, the lower ones usually a little earlier than the upper. The eyes open at about 14 days, but it takes a day or two until they are kept open permanently (Plate 9). About this time the

Harold Kalter   82

young begin eating solid food, and they can be safely weaned at about three weeks of age.

During the third week, the short, thick ears of the youngsters change rapidly into the large, thin ones of the adult. By the end of the third week, the young mouse completely resembles the adult, except for size and sexual maturity.

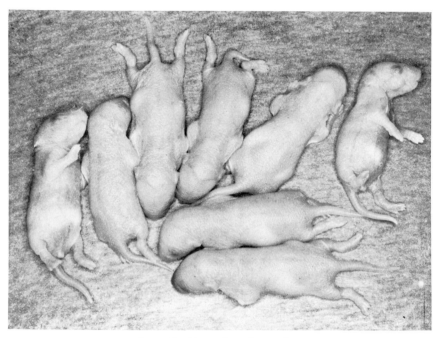

PLATE 2.—Litter of newborn mice.

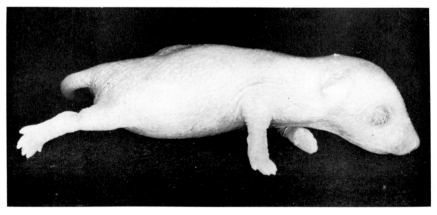

PLATE 3.—Newborn mouse. Note hairlessness, closed eyelids with pigment showing through, and pinna folded forward and down.

Maintenance of Inbred Colony

The order of these events is constant, but the time scale is variable and depends on a number of influences, especially birth weight; the larger the newborn, the more rapidly these changes occur.

PUBERTY

In young females the vagina is closed by a membrane. The age of vaginal opening, which marks the first estrus, varies considerably both within and

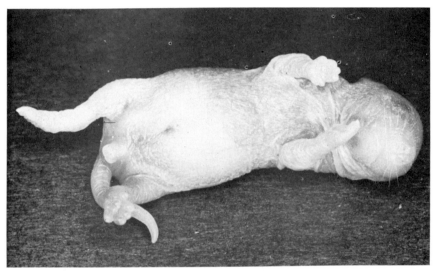

PLATE 4.—Newborn mouse. Note vibrissae, the only hairs present at birth, and genital tubercle.

PLATE 5.—Two-day-old mice. Note darkening of skin due to appearance of pigmented hairs within skin.

Harold Kalter 84

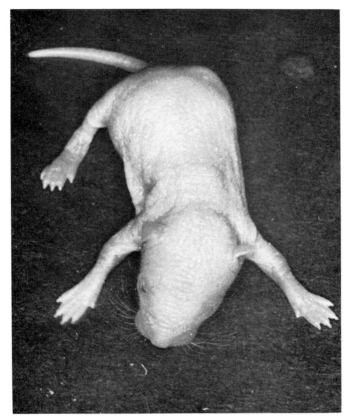

PLATE 6.—Three-day-old mouse. Left pinna is detached from skin of face.

PLATE 7.—One-week-old mouse. Ears are detached but thick and fleshy. Eyelids closed.

PLATE 8.—Eleven-day-old mouse. Baby coat full-grown. Eyes still closed.

PLATE 9.—Two-week-old mouse. Eyes still not fully open.

between stocks, but in general occurs at about 30 to 45 days of age. The first estrus is rarely a fertile one, however, and usually it is not until they are 7–10 weeks old that they first conceive. Maturity in males occurs at about the same time or a few days later.

Life Span

The useful breeding period of most females ends at about 10 to 12 months of age. Litters may continue to be produced after this time, but the estrous cycles and breeding behavior are irregular and litters small. Males usually breed for several months longer than females of the same stock.

Though mice may live as long as 3 years or more, the life span of most is much less, and by 16 to 18 months, 50 per cent are dead.

This has been a preliminary introduction to reproductive and growth phenomena of the mouse. Now let's find out what inbreeding is and what it does.

Inbreeding is, of course, the breeding together of genetically closely related individuals. For mice the most commonly practiced system is breeding of full brothers and sisters, but sometimes also of parents and offspring. The purpose of inbreeding is to produce genetic homogeneity in a group of animals and in this way to eliminate or reduce biological variability whose basis is genetic.

Of course, it is probably not possible to produce complete genetic uniformity, regardless of how long inbreeding is carried on, because of continually recurring mutations and other reasons. But for all practical purposes, it is considered that almost complete uniformity is achieved after about twenty generations of brother by sister matings. This takes about 4–5 years, and when this is accomplished, the animals are known as a strain; none but inbred animals should be called by this term.

Fortunately none of us has to go through the effort and time of producing inbred strains ourselves since inbred strains are commercially available. And this is a good thing, too, because it makes it possible for many different laboratories to use the same strains. Many local animal supply establishments now carry a few inbred strains and more are doing so all the time (1). The largest number of different strains and special types of mice is obtainable from the Jackson Laboratory in Bar Harbor, Maine. Its Production Department puts out a monthly *Supply Bulletin* which lists the current inventory.

A few other publications may be of interest and help. The first concerns the standardized nomenclature that has been adopted for use with inbred strains of mice (2). Next is a worldwide compilation of holders of inbred mice and the strains held, their characteristics, and other data. This is revised every two years. In addition the *Mouse News Letter* comes out twice a year. This contains a list of mutant genes and the laboratories that have them as well as news concerning studies with mice. These last two publications are sent to contributors. Inquiries about them may be sent to the Librarian, Jackson Laboratory, Bar Harbor, Maine.

The common practice of identifying rats by painting different colored spots on the back is not useful with mice, for several reasons: they're too small, you usually use many of them at a time, and often they are not albinos and so the spots are not easy to see.

Mice are best identified by punching holes and notches in the ears. An ordinary poultry punch can be used, but I have found those made by the Pettey Machine Shop, Hollywood, Florida, to be excellent. The left ear is used for the 10's and the right for the one's and in this way the animals can be numbered from 1 to 99 (Fig. 1). Sometimes the punched holes tear out at the edge of the ear, but the distinction between the holes and notches remains recognizable because the notches are shallow in comparison with the pulled holes. When 99 is reached you skip to 101, which is marked 01, of course. Ordinarily no animals with the same ear marks will be in the same cage, so commonly there is no opportunity for two mice with different numbers but the same ear marks to be mixed up.

Another practical matter is the method of keeping records. One of the principal purposes of keeping records of an inbred mouse colony is to be able to trace easily the ancestors, collaterals, and descendants of any particular animal. The method should also allow for recording the necessary biographical details of an animal or litter, and should be adaptable for recording experimental data. The methods of keeping such records vary widely in details, depending on the size of the colony and the particular needs of the investigator, but all systems are similar in certain features. An account of record keeping at a large genetics laboratory is given by Dickie (3).

My system is more or less the system used by Fraser, in whose laboratory I was introduced to it. This system makes use of ledgers and file cards. One ledger is used to record pregnancies serially, a second one is for recording pedigrees, and 3 × 5 cards are used for describing breeding pens.

Let's follow the routine in the animal room and see how, at every step, records are kept and what is recorded. We will begin with the breeding cages. These contain the breeding stock, the animals that provide not only the next generation of breeding stock but also a supply of experimental animals.

Each breeding cage has its number written on it. I use felt-nibbed pens for writing directly on the cage, a procedure I find more convenient than using cards attached to the cage. Each cage contains one to two males and three to five of their sisters. The cages are shoe-box size, and at the moment I have thirty such breeding units, but the number can be easily increased if neces-

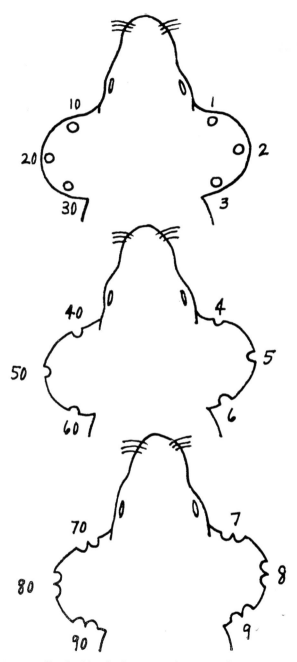

Fig. 1.—Numbering system by ear marks.

sary. For each cage there is a file card, stored in a single-unit file box, which is kept in the animal room. Figure 2 is a sample card. In the upper right corner is the cage number and just below it the date the animals were placed in the cage. In the left corner, inside the lines, is the number of the pregnancy that produced the litter from which the occupants of the cage came. This permits the rapid tracing of their ancestry. Also in the left corner, below the pregnancy number, is the cage number their parents occupied, and to the right, their birth dates. In the upper center is a symbol indicating the number of generations of inbreeding. "F?" means that I don't know for how many generations they were inbred before I received them, and "17" means that the

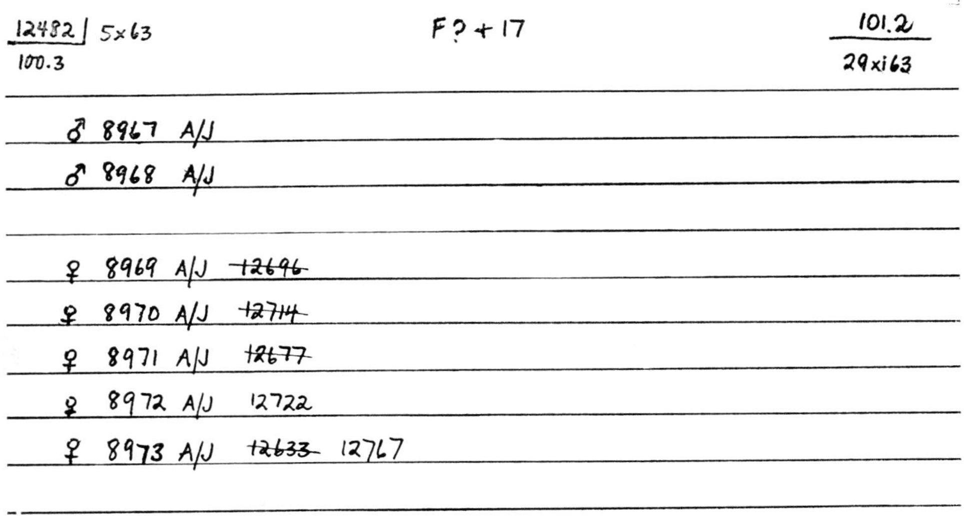

| 12482 | 5×63 | F? + 17 | 101.2 |
| 100.3 | | | 29×i63 |

♂ 8967 A/J
♂ 8968 A/J

♀ 8969 A/J ~~12696~~
♀ 8970 A/J ~~12714~~
♀ 8971 A/J ~~12677~~
♀ 8972 A/J 12722
♀ 8973 A/J ~~12633~~ 12767

FIG. 2.—Breeding-cage card. See text for explanation.

offspring of these mice are the seventeenth generation inbred in my laboratory. Below the first line are marked the sex, the individual's number, and the strain of the mice in the cage.

The breeding cages are examined once or twice a week, and females that are advanced in pregnancy are removed and placed in separate cages. They are given the next pregnancy number, and this is recorded next to their number on the file card, as shown in Figure 2.

The pregnancy is recorded in a ledger. Table 2 shows sample entries from this book. One line running across two pages is devoted to a pregnancy. Beginning on the left, first comes the pregnancy number. We will come back later to the next two spaces. Next is the number of the pregnant female, the number of the male or males, the breeding cage number, and the date of isolation. This much is recorded at the time a pregnant female is isolated.

TABLE 2

SAMPLE LEDGER

1	2	3	4	5	6	7	8	9	10
12722	......	unt	8972A/J	8937-8A/J	101.2	1i64	7i64	8	4♀, 4♂ 1♂unil(l)CLP, others out +0/7cp+1CLP 0,2disc, 4,2raised-7i64/4,2W-28i64
12723	......	unt	8743CBA	8737CBA	128.3	1i	5i	6,1d	2♀, 4♂(1d) all out+0/6cp 0,2disc,2,2raised-5i64/ 2,2W-26i64
12724	......	unt	8610DBA	8609DBA	140.2	2i	6i	10,3d	6♀(2d), 4♂(1d) all out+ 0/10cp 2,2disc, 4,2raised-6i64/3,2W-27i64
12725	VP	4×2.5Cort(11)	8817C3H	8632C3H	12463	5i	......	......	sacr 6+2res 4/6cp disc-22i64
12726	VP	Z+G(90)9-13	8796DBA	8714DBA	12578	6i	......	......	sacr 7+3res 5/7cp,oe,synd, etc. to B-23i64

NOTES TO TABLE 2

Sample entries from the "Pregnancy Book." Column 1—The pregnancy number, entered serially, recorded also on the breeding cage card, as shown in Fig. 2; 2—This space is used to indicate that a pregnancy is vaginal plug (VP)-timed, otherwise, left blank; 3—Here a pregnancy is indicated as untreated (unt) or the treatment is recorded: e.g., 4×2.5 Cort(11) = four daily injections of 2.5 mg cortisone acetate beginning 11 days post-VP, $Z + G(90)9-13$ = riboflavin-deficient diet (Z) + galactoflavin (G), a riboflavin antagonist, at a level of 90 mg/kg, for 96 hours, beginning nine days post-VP; 4—The number of the father; 5—The number of the pregnant female's number; 6—The breeding or holding cage number from which the female came; 7—The date isolated or on which the VP was seen; 8—The delivery date, in the case of non-experimental pregnancies, left blank if the pregnancy is interrupted. The last two columns are on the facing page. 9—The number of young delivered and the number dead (d) if any; 10—Remarks, e.g., number of male and female offspring; whether sacrificed (sacr); number resorbed (res); spontaneous malformations, such as cleft lip and palate (CLP); brief remarks concerning induced defects, number raised, number weaned (W), how disposed of, whether discarded (disc) or put in Bouin's solution (B), etc.

This system, therefore, uses a dual numbering system, one for each individual mouse and another for each pregnancy or litter.

The pregnancy cages are examined every morning, and when the female has given birth, the line in the pregnancy book is continued, by entering the date of birth, the number born, whether any are dead, the number of males and females, and remarks about the condition of the newborn.

The mother is then left as undisturbed as possible to nurse her young. About 3 weeks later the young are weaned, and the number weaned is entered in the book, completing the line. The female is returned to her breed-

TABLE 3

SAMPLE RECORD OF PEDIGREES

1	2	3	4	5	6	7
9108–09	♂ ♂	DBA	8610DBA	8609DBA	12601 | 1 ii63 140.2	143.1
9110–11–12	♀ ♀	DBA				143.1
9113	♂	C3H	8875C3H	8874C3H	12622 | 6 ii63 184.2	181.3
9114–15–16	♀ ♀	C3H				181.3
9117–18	♂ ♂	A/J	8973A/J	8967–8A/J	12633 | 6 ii63 101.2	expt.
9119–20	♀ ♀	A/J				expt.
9121	♂	?pc, brn	8632pc/+, brn	8630pc/+, dil. brn	12638 | 9 ii63 130.2	131.3
9122	♂	″ dil. brn				131.3
9123	♀	″ albino				131.3
9124	♀	″ albino				131.3
9125	♀	″ brn				131.3

NOTES TO TABLE 3

Sample entries from the "Pedigree Book." Each entry between horizontal lines refers to one litter. Column 1—Number(s) of individual animal(s); 2—Sex; 3—Strain or description; 4—Female parent's number and strain; 5—Male parent's number and strain. On the facing page: 6—In the "box," the pregnancy number producing the litter; to its right the date of birth; beneath it the breeding cage number of the litter's parents; 7—The breeding cage to which the litter is assigned or whether it is to enter an experiment.

ing cage, and her pregnancy number is crossed out (Fig. 2), indicating that she is once more in the cage. This, then, is one set of records.

The second ledger is the record of pedigrees. At some time before sexual maturity, the young animals are given their numbers, and their ears are punched. These numbers and other facts are entered in the pedigree book. A sample page is shown in Table 3. Beginning at the left are the individual's number, either several per line or one to a line depending on whether individual records are necessary, remarks about appearance or strain, and the parental numbers; on the facing page appear the pregnancy number, breeding cage number, date of birth, and to which role the individual is assigned, whether to form a new breeding unit or to enter an experiment. This information is the same as appears in the left corner of the 3×5 card (Fig. 2), thus providing the continuity that enables tracing ancestry, relationships, and so on among different generations.

In addition to this sort of information, the pregnancy record (Table 2) is also used to record pregnancies timed by plug, which are used for experimental purposes. The fact is recorded that the pregnancy is VP-timed (vaginal plug-timed), and treatment and other information are also recorded here.

REFERENCES

1. *Laboratory Animals. II. Animals for Research,* Publication 907, Washington, D.C.: National Academy of Sciences–National Research Council, 1961.
2. SNELL, G. D., *et al.* Standardized nomenclature for inbred strains of mice: second listing. *Cancer Res.* **20:** 145, 1960.
3. BURDETTE, W. J. (ed.). *Methodology in Mammalian Genetics,* San Francisco: Holden-Day, Inc., 1963.

SELECTED BIBLIOGRAPHY

GRUNEBERG, H. *The genetics of the mouse.* The Hague: Martinus Nijhoff, 1952.
SNELL, G. D. (ed.). *Biology of the laboratory mouse.* New York: Dover Pub., 1941.

CHAPTER 4

GENERAL MECHANISMS
OF TERATOGENESIS

MEREDITH N. RUNNER

Concepts and methods used in the study of congenital deformity should strive to uncover general mechanisms that account for abnormal development. Studies of this nature have been reported by Fraser, Kalter, Landauer, Russell, Warkany, and Zwilling, to mention a few. Students of abnormal development must remember that the ubiquitous genes play an important role in congenital deformity; first, because highly predictable and major mutants can produce abnormality irrespective of ordinary environmental variations; second, because other types of mutants showing variable expressivity and/or penetrance can be influenced by environmental factors; and third, because the less well-understood multiple gene effect plays a predominant role interrelating intrinsic and extrinsic factors that bring about abnormal development. These so-called modifying genes are important because they establish thresholds that, if transcended, produce abnormalities. Modifying genes may enter the picture in many places. They may act through the maternal organism and affect liver or kidney enzymes of the mother producing physiological alterations of the early embryo. Under direction of the mother's genes, the maternal organism supplies the new individual with the bulk of its initial complement of enzymes and cellular inclusions. Minor modifying genes also set the stage for threshold reactions within embryos that may render specific structures vulnerable to environmental variations. Embryonic susceptibilities are clearly illustrated by the phenomenon of cell degeneration that plays the role of sculptor in morphogenesis (1). Exaggeration of degenerative processes, for example, results in rumplessness or taillessness in chickens.

Interaction of genotype and irradiation is illustrated in Table 1, taken from experiments of Dagg (2). Two strains of mice, BALB/c and C57BL, were given 150 r of X-irradiation. The young showed cleft palate or abnormal feet, i.e., pre-axial polydactyly. Note that the modal time for cleft palate occurred in BALB/c embryos 5 mm in length at 11 days of age and that the modal time for polydactyly occurred in C57BL embryos 4.2 mm long and 10 days old. The experiments showed (*a*) that genotypes responded differently to X-irradi-

Professor of Biology, The University of Colorado, Boulder, Colorado.

ation, (b) that responses were stage specific; i.e., embryos showed evanescent genetic susceptibility, and (c) that susceptible genotype, C57BL, showed a spontaneous incidence corresponding to the modal type of deformity. The effects of treatment, therefore, depended upon genetically sensitive thresholds and the developmental stage of the test animals.

Having been reminded of the significance of genetic interaction with extrinsic agents as an important causative factor in congenital deformity, we proceeded on the assumption that every detectable deformity must have been permitted by the genetics of the individual. Whenever abnormality can be made to occur in predictable fashion, interaction with the residual genotype is also involved. This point of view swings our focus to a study of nongenetic factors that interact with the residual genotype to produce abnormality.

TABLE 1

EFFECT OF 150 R OF X-IRRADIATION
ON MOUSE EMBRYOS
Per Cent of Abnormal Young

STAGE	CLEFT PALATE		ABNORMAL FEET	
	BALB/c	C57BL	BALB/c	C57BL
Control.........	0	0	0	1.5
3.2 mm.........	1 (10 days)	....	8 (10 days)	....
4.2 mm.........	11 (10.5 days)	6 (10 days)	5 (10.5 days)	38 (10 days)
5.0 mm.........	35 (11 days)	....	10 (11 days)	....
6.0 mm.........	....	1 (11 days)	....	6 (11 days)

Analyses of non-genetic factors that are concerned with mechanisms require clarification of the concept of specificity. Teratogenic agents have been considered to be non-specific; they may produce a type of stress to which the embryo succumbs with an abnormality. However, a concept of non-specific teratogenic agents is created by the observer, for embryos respond in very specific ways. Consequences of anoxia, irradiation, and trypan blue may occur in all cells of the embryo, but the biochemical mode of action may remain obscure since only specific areas of the embryo become permanently deranged. Irrespective of treatment, responses are highly repeatable, hence specific, for those cells that become morphologically abnormal. The reasons for the localized effects then become a significant problem from a causal mechanism point of view.

The concept that stress produces developmental variations has been related to homeostasis by Lerner. He argues that an embryo with adequate genetic insurance has the capability of regulating normal development in spite of prenatal stress. The phenomenon has been called canalization by

Waddington, who has made development analogous to a branching track system with the tracks being located in canyons or valleys. Stressing agents alter the pathway of development. Tendencies to scale the walls of the canyon meet equilibrating forces (walls of the canyon) that channel development back to the proper pathway. As has often been stated, occurrence of anomaly is in part a measure of inability of genetic and other regulatory mechanisms to overcome localized sensitivities of embryonic tissues.

Assessment of the effects of treated embryos has probably been handled differently by each worker in the field. This shows the multiplicity of the possible methods for analyzing results of teratogenic experiments. Although the possible morphological deviations due to treatments are infinite in number, an analysis must be scaled to realistic proportions. The inadequacies of surveying for a single morphological endpoint must be emphasized. Admittedly we have selected the skeleton for studying interactions between pairs of teratogenic treatments. Suffice to say, had we confined our attention to abnormal ribs, for example, we would have reached a conclusion quite different from that which would have been indicated by attention focused on sacralization of the last lumbar vertebrae. Neither interpretation would have been substantiated by observations made exclusively upon abnormalities of the thoracic vertebrae.

Experimental teratogenesis in our hands has invariably produced a syndrome. Irrespective of whether the study has a genetic, a biochemical, or a comparative teratological approach, embryological material needs to be studied as an organismal syndrome rather than as a single structure isolated from the rest of the embryonic or fetal organism.

One experimental approach is to study effects of two treatments that simultaneously influence morphogenesis. The concepts are (a) that both normal and abnormal morphogenesis are expressions of poorly understood biochemical differentiations and (b) that interactions of two or more treatments that influence morphogenesis can, a posteriori, suggest mechanisms of action of the individual treatments and/or regulatory processes that determine normal morphogenesis.

Combinations of treatments have been used on a number of occasions and have usually shown potentiation of or protection from teratogenic and subteratogenic treatments. Miller (3) reported that fasting on the ninth day in combination with administration of cortisone produced a significant increase in the frequency of cleft palate. Kalter (4) has reported that the incidence of cleft palate was increased by food restriction in conjunction with administration of cortisone. Smithberg has shown that the teratogenic effect of insulin in mice has been increased by the addition of nicotinamide to the experimental treatment. Woollam and colleagues (5, 6) reported that cortisone and thiouracil potentiated the teratogenic effects of vitamin A and that vitamin A, in turn, increased the frequency of deformities due to X-irradiation. Russell (7) has shown and Brent (8) has provided supplementary data suggesting that hypoxia has a protective effect on the teratogenicity of X-irradiation.

Recent publications by Landauer illustrate the sophisticated level that has been achieved in his laboratory for studying mechanisms of action of teratogenic drugs. Table 2 (from a paper by Landauer and Clark [9]) shows the effects of a glucose analogue, 2-deoxy-D-glucose (2-DG) on the chick embryo when given by itself and in conjunction with a teratogenic dose of insulin. Experiments A and B indicate that 2-deoxy-D-glucose produced a decreased hatchability, i.e., was toxic, without producing a significant level of teratogenicity. Experiments B and C show that supplementation of 2-deoxy-D-glucose with normal glucose alleviated the toxicity of the drug. Similarly experiments

TABLE 2

INTERACTION OF INSULIN AND THE GLUCOSE ANALOGUE 2-DEOXY-D-GLUCOSE

(Given after 96 hours of incubation)

TREATMENT	EXPERIMENT							
	A	B	C	D	E	F	G	H
2-DG, mg.	0.25	0.75	0.75	....	....	0.6	....	0.6
Glucose, mg.	....	....	5.0	....	5.0	....	....	....
Insulin, unit.	....	....	....	2.0	2.0	....	1.0	1.0
Per cent hatch-toxicity	73	33	64	40	55	54	62	7.2
Per cent normal-teratogenicity ...	97	99	98	60	66	99	91	67

From Landauer and Clark (7).

D and E show that glucose reduced the toxicity of insulin without notably affecting the teratogenicity of the drug. These experiments may be considered preliminary to F, G, and H that show the results of giving one unit of insulin, 0.6 mg of 2-deoxy-D-glucose, and finally combining the two treatments. These comparisons show that dual treatments produced a potentiation of effect and that 2-deoxy-D-glucose increased both the toxicity and teratogenicity of insulin.

Insulin had at least two effects on the 4-day embryo: (a) reduced its ability to survive and (b) reduced its ability to develop normally. The data showed that for sake of investigation mortality and teratogenicity can be separated. Two types of reactions within the embryo occur. Both 2-deoxy-D-glucose and insulin interfere with intracellular utilization of glucose and thus lead to teratogenesis. Additional experiments by Landauer showed that some teratogenic agents acted at one of two critical stages which are also critical for the effects of insulin. Other teratogenic agents act at both critical stages. Interaction of drugs on developing embryos has enabled Landauer to deduce biochemical mechanisms by which the drugs bring about malformations.

In our laboratory, we have attempted to study mechanisms by which congenital deformity is induced in mouse embryos (10, 11). Table 3 shows that teratogenesis results from fasting; that is, removing all sources of nutrient from the mother for a 24-hr period during the ninth day of gestation resulted in 24 per cent of the offspring having deformities of ribs and axial skeleton. The syndrome included (*a*) abnormal vertebral centra in the lower thoracic region, (*b*) abnormal centra in the middle lumbar region, (*c*) abnormal ribs in the anterior thoracic region, (*d*) cranioschisis, (*e*) accessory ribs on the first lumbar vertebra, and (*f*) reduction of the cervical neural arches. The table

TABLE 3

PROTECTION EXPERIMENT ASSOCIATED
WITH FASTING

Treatment and Source of Carbon Bonds	Per Cent Abnormal
Fast, 24 hours, 9th day	24
Fast, glucose	2
Fast, casein	2
Fast, isoleucine	9
Fast, corn oil	6
Fast, acetoacetate	11

From Runner (10).

TABLE 4

CONVERGENCE EXPERIMENTS
WITH FASTING

Treatments	Per Cent Abnormal
Fast, glucose	2
Fast, glucose, iodoacetate	35
Fast, glucose, 7 per cent oxygen	47
Fast, glucose, 9-m-PGA	33
Insulin	55
Fast, glucose, trypan blue	17
Fast, glucose, X-ray, 100 r	13

From Runner (10) and Runner and Dagg (11).

shows a partial list of the nutrients that were administered to the fasted mice and the resulting frequency of abnormal young. Fasting by itself produced a frequency of 24 per cent abnormal offspring. Supplements such as glucose, casein, amino acids, fats, and the ketone body, acetoacetate, reduced the frequency to between 2 and 11 per cent. An interpretation from this array of protective agents is that protection came about by way of making available carbon-bonded energy in the organic molecules.

The second series of experiments (Table 4) was concerned with convergence of a series of treatments that produced comparable abnormalities in the mouse embryo. A series of treatments—fasting, iodoacetate, insulin, low oxygen tension, folic acid analogue, 9-methyl-pteroylglutamic acid (9-m-PGA), trypan blue, and X-ray—all produced essentially the same syndrome when administered at selected levels (11). Five of the seven treatments can be hypothesized to have a common denominator, that is, interference with the citric acid cycle. Trypan blue and X-irradiation are less informative because the effective route of action of these treatments is unknown. Thus the protection experiment indicated that energy bonds protect from the fasting effect, and the convergent experiments have indicated that a variety of treatments appear to produce abnormality by influencing the citric acid cycle.

The third series we call addition–non-addition experiments. Five treatments, giving comparable abnormalities in the embryo, were combined with fasting. Table 5 is divided into two parts. The top shows three dual treatments that appear to have simply an additive effect. The last two treatments

TABLE 5

NON-ADDITION EXPERIMENTS ASSOCIATED WITH FASTING

TREATMENT (Single or multiple pathway)	PER CENT ABNORMAL		
	Fed ad libitum	Fasted	(Expected)
Control.................	2	24	....
7 per cent oxygen.......	47	75	(71)
Trypan blue...........	17	44	(41)
X-ray, 100 r..........	13	33	(37)
Iodoacetate...........	63	66	(87)
9-m-PGA.............	13	19	(37)

From Runner and Dagg (11).

iodoacetate and 9-methyl-folic acid were non-additive when combined with fasting. The non-additive effect appears to be highly significant insofar as it indicates that once having interfered with normal development, the pathway or mechanism did not respond to a second treatment. This suggests that a single pathway or mechanism is involved. The experiments that showed an additive effect of dual treatment are equally significant, but the interpretations are less concise. Where multiple factors impinge upon specific developmental events additively, the data suggest that subthreshold environmental insults can act cumulatively to produce deformity: a concept that has considerable clinical significance. The embryonic mechanism may be either (a) that a single pathway is involved for the two treatments but the second treatment acts as an additive dose and embryos respond as if the dose intensity had been stepped up or (b) the alternate interpretation, that more than one pathway is involved and dual treatments acting through separate pathways hit the embryo twice instead of once.

Recently Grabowski (12) has reported occurrence of edema in the chick

Meredith N. Runner   100

embryo following treatment with anoxia. He relates the teratogenic effect of anoxia to alteration in the sodium-potassium balance. Grabowski's finding perhaps dovetails with a concept of reduced oxidative metabolism as an underlying cause of deformity. Interference with the citric acid cycle and consequent reduction in available energy would have as one of its early effects interference with selective permeability of the cell membranes. Two additional points should be made. First, energy deficient cells undergo protein synthesis and divide at a reduced rate. Second, and more significant, if modification of differential permeability is a connecting link between an immediate effect of anoxia and atypical morphogenesis, we must still account for the high degree of specificity and repeatability of deformities in localized areas of the embryo. Grabowski's experiments have explored a universal phenomenon in analyses of development. Irrespective of whether one studies the mechanisms of induction, of differentiation, or of teratogenesis there exists a 30- to 48-hr silent or lag period between the stimulus (organizer, inductor, or teratogen) and normal or abnormal morphogenesis.

Experiments on combined treatments have been extended in our laboratory so that the six treatments shown in Table 5 are currently being studied in all possible combinations. We found that X-irradiation has actually interfered with the abnormalities of ribs that would have been expected to occur as a result of fasting, anoxia, or iodoacetate. Since the irradiation (130 r) by itself produced 27 per cent of young with abnormal vertebrae but practically no deformed ribs, it cannot be said that fasting, anoxia, or iodoacetate protected embryos from irradiation but the reverse must be the case. Irradiation has interfered with the effect of fasting, anoxia, and iodoacetate insofar as rib abnormalities are concerned. The data further showed a cumulative effect of dual treatments (X-irradiation with fasting, and X-irradiation with iodoacetate) on vertebral abnormalities. Within the same embryos therefore, combined treatments gave a concomitant additive or synergistic effect upon abnormalities of the vertebrae and an interference effect upon abnormalities of the ribs.

Interference of two teratogenic agents, especially showing that X-ray is protective, may be without precedence. Perhaps our knowledge about interaction of teratogenic agents has reached the point where we may be able to reverse the procedure and, by the common denominator approach, ask by what mechanisms do irradiation and trypan blue cause malformation.

How are these experiments concerned with the problem of drug safety? Obviously, the results obtained from mice may not be directly extrapolated to drug effects in human beings. I believe that we have shown that normal development in mouse embryos, and probably all mammalian embryos, is dependent on carbohydrate metabolism. We have opportunistically exploited a peculiarity of the mouse mother because of her inability to store sufficient carbohydrate to provide for an acute period of deficiency. Although we have interpreted these results as indication of energy starvation in embryonic tissue, such interpretations must be made cautiously, for in the words of Landauer (9), "the unanalyzed and surely complex reverberations which these treatments had in the maternal organism can scarcely be accepted as evidence that energy starvation per se is a factor in the origin of congenital malformation."

A perplexing problem today is how to determine whether drugs are safe during pregnancy. How can testing be done? I contend that the basic question should be how do drugs act? We know that there are differences between genotypes, between strains, between species, and between individuals. Superficial information obtained from mice may not directly apply to teratogenesis in man, but truly basic information should permit direct transfer of information from species to species. That we are unable to see a clear course to follow for testing is an expression of our lack of understanding of differences between mice and men. The fact that thalidomide works in man and perhaps not in the mouse could be a matter not of teratogenicity of the compound per se but a matter of how rapidly thalidomide is or is not catabolized by liver enzymes.

So I return to the dependence of the mouse embryo on available carbohydrate—a finding that I would expect to be applicable to, but not readily demonstrable, in animals more fortunately provided with stored carbohydrates. Experiments on interactions of teratogenic agents promise to serve as a double-edged sword. One, experiments show that there is interaction with potentiation and/or interference of multiple agents. The multiple-factor hypothesis is almost a foregone conclusion; otherwise more deformities in man would long ago have been correlated with specific causative agents. The other edge of the sword is that multiple experimental treatments enabled us to deduce, *a posteriori*, mechanisms by which normal embryos undergo differentiation.

In summary, then, the ubiquitous genome behaves in permissive fashion because it establishes the ease with which one or more thresholds can be transcended to produce a teratogenic syndrome. Dual teratogenic treatments have enabled investigators (e.g., Landauer) to dissect multiple consequences of a given treatment and by the common denominator approach have suggested biochemical mechanisms of normal and abnormal development.

The teratogenic effect of fasting in mice studied by protection, convergence, and non-addition experiments has implicated the citric acid cycle to be essential for normal development and has demonstrated localized sensitivities of the embryo—presumably to altered carbohydrate metabolism.

Interaction of treatments has shown that X-irradiation at a teratogenic level has actually interfered with the consequences of other teratogenic treatments and has protected specific areas of the mouse embryo from deformity. This approach promises to enlighten us about the mechanisms by which teratogens such as X-irradiation cause abnormal development.

Delineation of mechanisms for normal and abnormal development currently shows promise of major achievements. Permanent research and training centers, more extensive than now exist, are needed in order that we may focus recent spectacular advances in genetic coding, tissue culture, biochemistry, and biophysics upon the problem of teratogenesis. The interdisciplinary nature of teratology is a challenge to talented scientists from fields of genetics, embryology, pharmacology, behavior, biochemistry, and biophysics who wish to collaborate on broad and significant problems. The sophisticated techniques available and the elaborate facilities essential for today's research point to the need for a major national center in order to pursue an integrative approach to problems of normal and abnormal development.

REFERENCES

1. SAUNDERS, J. W., GASSELING, M. T., AND SAUNDERS, L. C. Cellular death and morphogenesis of the avian wing. *Dev. Biol.* **5**: 147, 1962.
2. DAGG, C. P. Some effects of X-irradiation on the development of inbred and hybrid mouse embryos. *Proceedings of the international symposium on the effects of ionizing radiation in the reproductive system.* Pergamon Press, 1963.
3. MILLER, J. R. A strain difference in response to teratogenic effect of maternal fasting in the house mouse. *Can. Jour. Genetics Cytol.* **4**: 69, 1962.
4. KALTER, H. Teratogenic action of a hypocaloric diet and small doses of cortisone. *Proc. Soc. Exp. Biol. Med.* **104**: 518, 1960.
5. WOOLLAM, D. H. M. AND MILLEN, J. W. Influence of 4-methyl-2-thiouracil on the teratogenic activity of hypervitaminosis A. *Nature* **181**: 992, 1958.
6. WOOLLAM, D. H. M., MILLER, M. A., AND FOZZARD, J. A. F. Influence of cortisone on the teratogenic activity of X-irradiation. *Brit. J. Radiol.* **32**: 47, 1959.
7. RUSSELL, L. B. *et al. Anat. Rec.* **111**: 455, 1954; *Radiation Biol.* **904**: 5, 1954; *Suppl. Jour. Cell. Comp. Physiol.* **43**: 129, 1951.
8. BRENT, R. L., FRANKLIN, J. B., AND BOLDEN, B. T. Modification of irradiation effects on rat embryos by uterine vascular clamping. *Radiation Research* **18**: 58, 1963.
9. LANDAUER, W. AND CLARK, E. M. Teratogenic interaction of insulin and 2-deoxy-D-glucose in chick development. *Jour. Exp. Zool.* **151**: 245, 1962.
10. RUNNER, M. N. Inheritance of susceptibility to congenital deformity. Clues provided by experiments with teratogenic agents. *Pediatrics* **23**: 245, 1959.
11. RUNNER, M. N. AND DAGG, C. P. Metabolic mechanisms of teratogenic agents during morphogenesis. National Cancer Institute Monograph #2, 1960, pp. 41–54.
12. GRABOWSKI, C. T. Teratogenic significance of ionic fluid imbalances. *Science* **142**: 1064, 1963.

TRANSPLANTATION OF MAMMALIAN OVA: BLASTOCYST AND EARLIER STAGES

Purpose

To present the technique of transplanting fertilized and unfertilized ova for prenatal foster mothering.

Materials

Female mice within 3 days postcoitum; zygotes with appropriate genetic markers for accurate identification; watchmaker's forceps; fine scissors for the skin incision; 2-inch watch glasses with ring support; hypodermic syringe for flushing the uterine lumen; 70 per cent Locke's solution for a suspension medium; special capillary pipette for transferring zygotes; needle and thread for closing the skin incision; dissecting microscope with substage mirror and illumination; anesthesia ether or veterinary Nembutal; 75 per cent alcohol for swabbing the hair and skin before making incision.

Materials for making the special glass pipette consist of a Bunsen burner with a wing tip for drawing the capillary tubing, Dekhotinsky cement for inserting the capillary tubing into the holder, 0.25-inch rubber tubing and mouth piece, and a fine carborundum stone for cutting the capillary tubing.

Method

Treatment *in vitro* and transplantation of zygotes offer opportunity to study environmental influences upon the early zygotes under well-controlled conditions and independent of possible interactions with the maternal organism. The procedures can be divided into (*1*) preparation of the donor and host animals, (*2*) the recovery operation, (*3*) suspension medium and treatment procedures, (*4*) introduction into the host reproductive tract, and (*5*) assessment of results.

Preparation of donor and host animals varies somewhat, depending upon whether transplantations and treatments will be done at, or just before, sperm penetration, at cleavage stages, or at the blastocyst stage prior to attachment to the uterus. The procedures differ mainly in timing of the operation in relation to copulation, to the site of recovery, and to the site of implantation of the zygotes. Workers seem to agree that the highest per cent of success is attained by transplanting 3-day blastocysts into the recipient uterus 2 days postcoitum. Eggs prior to or just about the time of fertilization are recovered from the oviduct and are transplanted into the ovarian bursa where they are picked up by the recipient oviduct.

In all experiments it is essential that the genotypes of the transplanted and native embryos are different so that unequivocal distinctions can be made between transplanted and native offspring. Eye pigmentation in the mouse appears at 12.5 days postcoitum so that if transplanted and native embryos differ by having pigmented and non-pigmented eyes, the diagnosis of success of transplanted embryos can be established at that time. This genetic tag is especially convenient because the pigmented and non-pigmented eyes can be distinguished at birth whereas coat color cannot be determined for 3 to 10 days postpartum.

The most dependable donors in our laboratory are obtained by selecting from a pool of adults those females that by vaginal inspection are judged to be in estrus. This must be determined in the evening or in the early part of the dark period for a colony having artificial light cycles. Estrous females are introduced singly into a pen containing an experienced male. Examination for the vaginal plug is made about 3 hours later. Penetration of ova by sperm occurs within 2 hours. Cleavage into a 2-celled zygote occurs about 24 hours later. Cavitation of the morula begins about 48 hours postcoitum, and the blastocyst can be recovered from the uterus about 72 hours after mating. If unfertilized eggs are needed in large quantity these can be obtained most easily following hormonally induced ovulations at about 25 days of age. The procedure is shown in Figure 1 and consists of a priming dose of pregnant mare serum followed 34 hours later by an ovulatory injection of luteinizing hormone or, conveniently, chorionic gonadotropic hormone. Ovulation begins about 10 hours following the ovulatory injection and is essentially completed by 15 or 16 hours after the injection (Fig. 2). Using this procedure we have recovered over 100 eggs from exceptional animals.

Unfertilized eggs, enclosed in follicular cells, can be seen through the distended wall of the ampulla. The section of the oviduct is excised and put into a watch glass half-filled with suspension medium. A nick in the wall of the ampulla with fine-pointed watchmaker's forceps allows muscular tone of the oviduct to force the mass of ova through the tear in the wall. Recovery of 2-celled and morula stages from the oviduct is attained by excising the entire oviduct and placing it in the watch glass. Mincing the oviduct in short segments with fine scissors again enables the spontaneous muscular contractions to deliver the contained ova. The mascerated pieces of oviduct are removed and the solution allowed to stand for a few seconds. The ova in their spherical zona pellucida roll to the deepest part of the watch glass and can be detected by the smooth, highly refractile zona pellucida. Recovery of blastocysts from the uterus is achieved by transecting at the cervical end and removing the entire uterus including the uterotubal junction. Excess blood is removed by blotting or rinsing, and the uterus is put into a watch glass with the suspending medium. The tips of the uterine horns are clipped off with

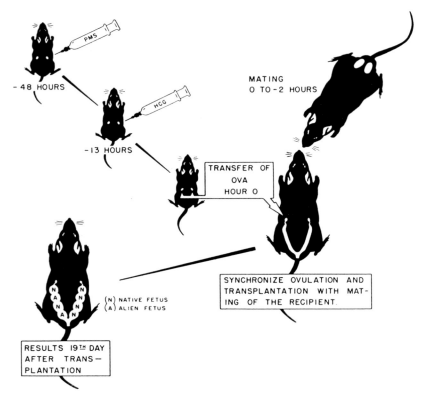

FIG. 1.—Transplantation of superovulated ova

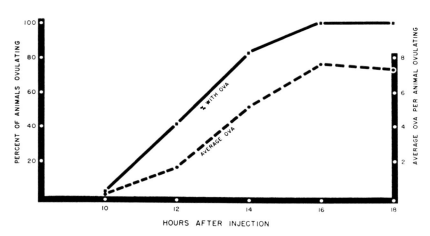

FIG. 2.—Time of induced ovulations in mice

fine scissors. A syringe, equipped with a Number 27 needle and filled with suspension medium, is used to flush out the uterine contents. The needle is inserted through the cervical opening and both needle and uterus are grasped with the forceps. The needle is inserted first into one horn where about half of a milliliter of the solution is expelled to irrigate the uterine horn. Then the needle is inserted into the other uterine horn to irrigate it with another half of a milliliter of the solution. The uterus is discarded, the solution allowed to settle, and the zygotes with their glistening zona pellucida can be seen at the bottom of the watch glass. Blastocysts are relatively resistant to ordinary environmental manipulations as has been demonstrated by the fact that they may be held for several hours. Chang (1) has shown that rabbit blastocysts can be stored in the refrigerator for several days or held for subsequent development. Lin *et al.* (2) have shown that mouse blastocysts can be subjected to subfreezing temperatures and retain capability to survive. Both Chang and Hunt (3) and Glass and Lin (4) have subjected unfertilized eggs to irradiation to test their ability to complete subsequent development.

Citrate-buffered egg yolk medium for mouse ova:

Sodium chloride	600 mg
Sodium citrate	30 mg
Potassium chloride	30 mg
Calcium chloride (anhyd.)	15 mg
Magnesium chloride	3 mg
Glucose	100 mg
Water (to make)	95 ml
Egg yolk	5 ml

(Centrifuge before using.)

Suspension media for mammalian ova have varied from Locke's saline solution to complex culture media. A number of workers using laboratory animals have cultured mouse eggs for the 2-celled embryo to the blastocyst and subsequently have tested the capability of cultured embryos to develop *in utero*. There is an apparent sensitivity of mouse ova at the 1-cell stage. A series of experiments in our hands testing media for transplanting unfertilized eggs indicated relatively little difference among media such as 70 per cent Locke's solution, Locke's solution with yolk-citrate buffer, and a more complex culture medium designed by Earle (5). Apparently, factors other than the medium influenced the capability of development more than did the media being tested.

Preparation of recipient animals consists primarily of establishing a suitable stage of pseudopregnancy. As indicated above, close synchrony of donor and recipient has been regarded as advantageous for unfertilized ova. Asynchrony consisting of approximately a 24-hour reduction in postcoital time for the recipient animal seems advantageous for transplantation of morulae and blastocysts (6).

Choice of anesthesia for the recipient seems to be a matter of convenience for the operator. Experienced operators in our laboratory invariably prefer ether, although attention is required during the course of the operation. The advantage seems to be that full depth of anesthesia is attained almost immediately and the length of the anesthesia can be adjusted to fit the operation. Prompt return to consciousness seems to be advantageous for the mouse, and a short period of care for the recipient animal offers an advantage to the operator. Barbiturates, for example, veterinary Nembutal, are convenient because they produce a deep anesthesia (so deep in fact that the body temperature ordinarily drops) and attention is not required during the course of the operation. The disadvantage is that the injection must be given approximately 10 minutes before the operation and the animals are not ambulatory when the operation is finished. Veterinary Nembutal for use with mice is prepared by withdrawing 0.25 cc. in a hypodermic syringe and diluting this with saline, 25 cc. The stock solution is injected intraperitoneally at a dose of 0.01 cc/g of body weight. For example, a mouse weighing 20 g is given 0.2 cc (5 mg/kg) of the solution.

The area for the incision to expose the oviduct or uterus is wetted with alcohol and the hair is parted for entry into the peritoneal cavity just posterior to the last rib. A skin incision, about 0.5-inch long, is made with the fine scissors and the body wall is penetrated by a blunt dissection with the forceps. The anterior part of the reproductive tract is delivered through the incision and care is taken not to use the forceps on any part of the oviduct or uterine wall. The ovarian fat pad and the mesovarium or mesometrium provide useful handles for manipulating the reproductive tract.

Transplantation of ova or recently fertilized zygotes to the reproductive tract of the recipient is made by means of a capillary pipette. The act of transplanting to the ovarian capsule is illustrated in Plate 1 taken from Runner and Gates (7). The fat pad provides a convenient anchor for the forceps, and the tip of the pipette can be inserted into the fat and through the ovarian capsule. The ova with their follicular cells are easily seen as they leave the tip of the pipette and enter the ovarian bursa. Inexperienced operators find it convenient to stain the ova with methylene blue for easier visibility during the course of the transplantation. To the best of our knowledge the stain has no deleterious effect on the ability of the ova to survive. Transplantation of morulae and blastocysts to the uterine lumen differs in only one minor respect. The uterus is held by the mesentary (near the uterotubal junction) and a fine needle puncture is made at the anterior tip of the uterine horn. The capillary pipette is promptly inserted into the puncture, and the zygotes are expelled with care so that air is not added. A convenient technique for assuring delivery of the zygotes without expelling air into the uterine lumen is to let the capillary tube fill with the suspending medium until it reaches an equilib-

rium. Then a very short column of air is drawn in the pipette and the blastocysts are drawn below the marker bubble. When the contents of the pipette
are expelled into the uterine lumen, the bubble can be seen and followed
more easily than the eggs being transplanted. The pipette is withdrawn, and
the punctured surface is held together for an instant with a fine pair of
forceps before returning the reproductive tract to the peritoneal cavity. We
find that a minimal incision through the body wall requires no suture and
two stitches or metal clips are put into the skin incision.

Recipients can be prepared by mating them with vasectomized males so
that in fact they are pseudopregnant until the time that fertilized eggs are

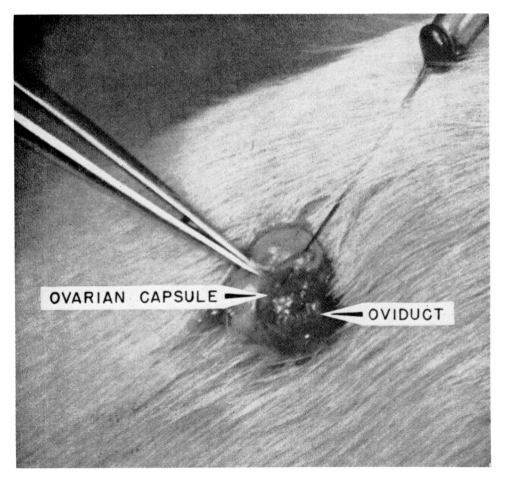

PLATE 1.—The method used for transplanting unfertilized eggs is illustrated above. Egg production was induced in a normally sterile obese female. The eggs were transferred to a glass pipette
and injected into the ovarian capsule of a non-obese foster mother. The transplanted eggs were
gestated along with the foster mother's own progeny.

provided. This procedure has at least two disadvantages from our point of view: (*1*) several cases of spontaneous repair of the vas deferens has shaken our confidence in use of so-called sterile males, and (*2*) more important, we prefer to work with animals that are in fact pregnant, thereby providing a control uterine horn for the experimental side. In fact, we regard it to be an experimental advantage to have the transplanted embryos intermixed with native offspring.

It might be appropriate to point out that, among hundreds of the offspring that have been seen from the transplantation experiments, none of the investigators has obtained the impression that the technical manipulation accounts for increased frequency of malformations. These offspring are notably free of malformations—certainly as free as are the controls.

Some experimenters have suggested the possibility that preimplanted embryos would make convenient test objects, independent of maternal complications, to test for teratogenicity. It is unscientific to discourage inquiring points of view; therefore, biased opinions must be held in check when others have designed experiments to test hypotheses. In this instance, however, good evidence exists to show the futility of attempting to consistently induce malformation by treating embryos prior to implantation. Experiments with irradiation on mammalian material (rabbits, rats, mice, and hamsters) by workers such as Russell, Wilson, Brent, and Harvey (recently reviewed by Inman and Marikvee (8) have shown resistance of mammalian embryos to congenital deformity prior to formation of primary germ layers. Additional evidence with the lower vertebrates—the best evidence coming from amphibian material (enucleated eggs, eggs damaged in many ways, hybridization experiments, and transplantation of nuclei)—has shown that embryonic development is autonomous through the blastula and early gastrula stages. Morphological responses to insults appear from treatments administered after the primitive streak stage. Evidence from many laboratories indicates that the genome first makes its influence known about the time that the third germ layer appears. This apparently is the time that the first significant effects of information from nuclei become integrated into a series of morphological events. The coincidence of regulation of morphology by the genome and the abrupt onset of sensitivity to teratogenic agents may be of more than passing significance.

REFERENCES

1. CHANG, M. C. Storage of unfertilized rabbit ova: Subsequent fertilization and the probability of normal development. *Nature* **172:** 353, 1953.
2. LIN, T. P., SHERMAN, F. K., AND WILLETT, E. L. Survival of unfertilized mouse eggs in media containing glycerol and glycine. *Jour Exp. Zool.* **134:** 257, 1957.
3. CHANG, M. C. AND HUNT, D. M. Effects of *in vitro* radiocobalt irradiation of rabbit ova

on subsequent development *in vivo* with special reference to irradiation of maternal organisms. *Anat. Rec.* **137**: 511, 1960.

4. Glass, L. E. and Lin, T. P. Development of X-irradiated and non-irradiated mouse oöcytes transplanted to X-irradiated and non-irradiated recipient females. *Jour. Cell. Comp. Physiol.* **61**: 53, 1963.
5. Smithberg, M. and Runner, M. N. A suitable medium for use with transplantation of ova in the mouse. *Anat. Rec.* **115**: 414, 1953.
6. Doyle, L. L., Gates, A. H., Noyes, R. W., and Nauber, M. Asynchronous transfer of mouse ova. *Fertility and Sterility,* in press.
7. Runner, M. N. and Gates, A. Sterile, obese mothers. *Jour. Hered.* **45**: 51, 1954.
8. Inman, O. R. and Marikvee, C. R. Gross effects on rabbit embryos and membranes of X-irradiation in the blastocyst stage. *Anat. Rec.* **147**: 139, 1963.

Selected Bibliography

McLaren, A. and Biggers, J. D. Successful development and birth of mice cultivated *in vitro* as early embryos. *Nature* **182**: 877, 1958.

Runner, M. N. and Gates, A. Conception in prepuberal mice following artifically induced ovulation and mating. *Nature* **174**: 222, 1954.

Runner, M. N. and Palm, J. Transplantation and survival of unfertilized ova of the mouse in relation to postovulatory age. *Jour. Exp. Zool.* **124**: 303, 1953.

CHAPTER 5

NUTRITIONAL FACTORS IN MAMMALIAN TERATOLOGY

E. MARSHALL JOHNSON

This discussion will be limited to studies of teratogenic vitamin deficiencies in the rat. Emphasis will be placed on the timing of the deficiency in relation to embryonic development and on pathogenesis of the malformations. The teratogenic effects of only five vitamin deficiencies have been studied extensively. These five vitamins are vitamin A, vitamin E, riboflavin, pantothenic acid, and pteroylglutamic acid. To the last-named vitamin must be added some studies incorporating B_{12} deficiency as well. Recently, niacin deficiency has come under renewed investigation (1, 2) through the use of the antagonist, 6-aminonicotinamide; but the relatively few reports on production of congenital abnormalities in the rat from deficiencies of choline, methionine, thiamine, vitamin D, vitamin K, and the unsaturated fatty acids require additional study. For information relating to these and other teratogenic procedures, see the comprehensive review by Kalter and Warkany (3).

The experiments to be discussed can be divided into three main types:

1. A chronic deficiency in which the mother is given a diet containing ample amounts of all but one essential nutrient. The nutritional factor to be studied is fed in low levels initially to permit breeding and is then deleted throughout pregnancy.

2. An acute deficiency which is similar to the above but is terminated by supplementation at some point during pregnancy.

3. A transitory deficiency which differs from an acute deficiency in that the mother is fed a complete ration during the first and last portions of the pregnancy and is deficient in a single nutrient for only a short period during pregnancy. Such transitory deficiencies usually are achieved by the use of an antimetabolite during the second week of the 3-week gestation of the rat.

VITAMIN A

Warkany and Schraffenberger (4) reported the production of congenital malformations of the eyes and later (5) of the soft tissues with a mild, chronic

---

Assistant Professor of Anatomy, College of Medicine, The University of Florida, Gainesville, Florida.

vitamin A deficiency. Weanling rats were raised to maturity on a diet low in vitamin A, i.e., a vitamin A-deficient diet supplemented with low levels of carotene. Then these rats were bred and maintained on a diet deficient in the vitamin throughout pregnancy. The incidence of congenital malformations or embryonic death varied with the carotene or vitamin A supplementation used before breeding. By these means, the dividing line between abnormal embryonic development and fetal death was demonstrated to be quite narrow. That is, an increase in the carotene supplement resulted in increased fertility and decreased percentage of abnormal young. Reduction of the carotene supplement decreased the number of young per litter but increased the percentage of young showing congenital malformations. The observed malformations of the soft tissues included those of the genitourinary tract and of the cardiovascular system as well as diaphragmatic hernia, edema, subcutaneous hemorrhage, and underdevelopment of the lungs.

Oral supplementation with a single dose of vitamin A (16,000 International Units) during the second week of pregnancy altered the incidence and types of anomalies encountered at autopsy as well as the incidence of embryonic death (6). Such vitamin supplementation on the ninth or tenth day[1] of pregnancy prevented abnormalities of the aortic arches, the lungs, the diaphragm, and the genital ducts. Anomalies of the latter two organs were reduced in incidence by supplementation as late as day 14. The usual ocular anomalies could be prevented by supplementing the mother's diet as late as the eleventh day and were reduced in incidence by dosage as late as day 14. However, previously unobserved abnormalities of the eye occurred even with supplementation as early as the ninth day. Similar findings were reported in regard to the usually observed abnormalities of the kidneys, ureters, and gonads. On the other hand, hydronephrosis and hydroureter, defects previously unobserved, occurred even after supplementation as early as the ninth day of pregnancy. Abnormalities of the lower genitourinary tract were virtually prevented by supplementation as early as day 9 or as late as day 14. In contrast, anomalies of the heart could not be prevented by supplementation even on the ninth day. From these findings they concluded (6) that "the malformations resulting from maternal vitamin A deficiency were determined during the period of active organ formation. . . ."

Malformations of the eyes occurred frequently in the young from vitamin A-deficient mothers (7, 8). One effect observable grossly was failure of normal fusion of the eyelids, resulting in "open" eyes. Histological examination revealed retinal foldings, retinal coloboma, and eversion as well as anomalies of the iris. However, the most frequently observed abnormality was prolifera-

---

[1] The timing of days of pregnancy given in reports from other laboratories has been changed to correspond with standards of timing in which the day of finding spermatozoa in the vagina is considered to be day 0 of pregnancy.

tion of connective tissue in the vitreous chamber. A detailed analysis of the cardiovascular malformations in these abnormal young, ranging from 16 to 21 days fetal age, was made by Wilson and Warkany (9, 10). Aortic arch anomalies occurred in 34 per cent, interventricular septal defects in 34 per cent, and general retardation of myocardial development in 50 per cent of the specimens. It was concluded that all malformations were the result of interference with normal processes occurring on or subsequent to the eleventh day of pregnancy. In addition, anomalies were found to be more frequent in the younger fetuses, indicating that these defects may have been associated with intra-uterine death and resorption.

In detailed histological studies in young from more than 100 litters, Wilson and Warkany (11, 12) observed many anomalies of the genitourinary tract and, in addition, keratinizing metaplasia of the epithelium derived from the urogenital sinus. This keratinization may be considered evidence of a fetal vitamin A deficiency as similar changes result from vitamin A deficiency in adult rats. It was concluded that abnormalities of the genitourinary system (other than keratinization) and of other systems were the result of paraplasia, hypoplasia, or aplasia. The only instance of paraplasia was felt to be "fused" or "horseshoe" kidneys and perhaps a homologous set of genital ducts. Many instances of hypoplasia were described, e.g., late or faulty partitioning of the cloaca and urogenital sinus, retarded Müllerian duct development, renal ectopia, ectopic ureteric opening, hypospadias, cryptorchidism, pulmonary and renal hypoplasia, delay in fetal trabeculation and myocardial development, and diaphragmatic hernia. Wilson and Warkany (12) believed that they had observed true aplasia of the male accessory glands of reproduction and the vagina.

Vitamin E

That the effects of vitamin E deficiency on embryonic development lead to fetal death and resorption has long been known (13). Evans *et al.* (13) and Mason (14) concluded that the primary effect of the deficiency was on embryonic mesenchyme. However, the first report of frank congenital malformations resulting from this vitamin deficiency was that of Thomas and Cheng (15). These investigations reported that vitamin E-deficient rats would produce normal young when given a single oral supplement of the vitamin (1.2 mg) as late as the eighth day of pregnancy. Without supplementation, all embryos died and were resorbed, but when the supplement was given on the ninth, tenth, eleventh, or twelfth days, living young with abnormalities of diverse types were observed. These anomalies included hydrocephalus and exencephalus, anophthalmia or microphthalmia, umbilical herniations and ectocardia, cleft palate and harelip, and skeletal defects. When the vitamin supplement was administered on the ninth day, hydrocephalus was the most common anomaly. When the vitamin supplement was administered on the tenth day, umbilical herniations resulted and on day 11 both exencephalus and umbilical herniations were frequent (16). The highest incidence of abnormalities (62 per cent) occurred after vitamin supplementation on the tenth day.

In a more detailed study (17) the pregnant vitamin E-deficient rats were given 2 mg of alpha-tocopherol acetate on the tenth day, and groups of rats were sacrificed each day until term to allow a chronological study of abnormal development. On the eleventh day the embryos were grossly reduced in size and retarded in development when compared with control embryos, and by the thirteenth day they showed grossly detectable signs of necrosis and resorption. In agreement with Evans *et al.* (13) and Mason (14), Cheng and co-workers (17) concluded that mesoderm appeared to be the germ layer that was most severely affected by vitamin E deficiency under these conditions.

RIBOFLAVIN

Between 1940 and 1944, Warkany and associates published an extensive series of papers describing the production of young with skeletal abnormalities and demonstrated that these defects resulted from a chronic deficiency of riboflavin. At first the Steenbock and Black rachitogenic diet supplemented with vitamin D was used. Rats maintained on this diet bore young with a variety of skeletal defects such as short mandible, cleft palate, syndactylism, short tail and legs, fused ribs and sternebrae, and reduction or absence of ossification (18, 19). When supplementation with liver or liver extract was begun on the twelfth day of pregnancy, the occurrence of skeletal abnormalities in the young was prevented, whereas supplementation after day 13 did not prevent the skeletal defects (20). It was concluded that the developmental period between the twelfth and thirteenth days was the critical period for skeletal malformations resulting from maternal riboflavin deficiency. With a purified diet lacking in riboflavin, young with identical skeletal defects were produced and these defects were prevented by crystalline riboflavin given throughout pregnancy (4, 7, 21).

Histologic studies by Warkany and Nelson (22) showed a marked delay in ossification of the grossly abnormal bones. That is, the normal relationship between the calcifying cartilage and osseous portions of the bone was deranged in that cartilage persisted in areas that should have been ossified. Moreover, the cartilage models themselves were abnormal prior to the beginning of endochondral ossification, and the radius, ulna, tibia, and fibula were often shortened in the cartilaginous stage of development. Abnormal position of some ossification centers was also observed. Syndactylism and brachydactylism were considered to result from a failure of longitudinal or transverse division in the cartilaginous models of the paws (23). It was concluded that the defects in ossification were secondary to faulty developmental processes in the cartilaginous and precartilaginous stages. The variation in susceptibility of different bones to riboflavin deficiency has previously been emphasized (19).

In a detailed study of the palate-facial area, Warkany and associates (24, 25) reported that cleft palate resulted from retarded development and faulty position in the palatine processes. Tooth development was often markedly retarded, especially in the more anterior portion of the dentition. The mandible was short with most of the reduction in size being due to hypoplasia of the body posterior to the premolar area.

The production of young rats with skeletal defects by a chronic maternal riboflavin deficiency has been confirmed by many laboratories, e.g., Giroud and Boisselot (26). An acute riboflavin deficiency during pregnancy has been produced by the use of the vitamin antimetabolite, galactoflavin, with a riboflavin-deficient diet (27). When this regimen was employed throughout pregnancy, a high incidence of skeletal anomalies with various anomalies of the soft tissues were observed. When the antimetabolite-containing diet was given from days 7 to 13 and followed thereafter by vitamin supplementation, the offspring exhibited principally skeletal and cardiac abnormalities. Limiting the deficiency period to days 7–11 resulted in only cardiovascular anomalies. This study demonstrated that riboflavin deficiency during pregnancy could affect not only the skeleton but also the cardiovascular and urogenital systems, the cerebrum and eyes, and the body walls. When diets containing the highest level of the antimetabolite were supplemented with riboflavin throughout gestation, fetal development was normal. Similar results have recently been obtained with galactoflavin-containing, riboflavin-deficient diets of several strains of mice (28), and strain differences in response to the teratogenic action of galactoflavin have been demonstrated.

Pantothenic Acid

Boisselot (29–31) demonstrated that instituting pantothenic acid deficiency 3 to 10 days before mating resulted in a high incidence of abnormalities in the few young still surviving until term. The malformations observed included exencephaly and pseudoencephaly, anophthalmia and microphthalmia, subcutaneous edema, and digital hemorrhages resulting in deformed paws. When the pantothenic acid-deficient diet was started 10 to 20 days before breeding, 10 μg of calcium pantothenate daily did not prevent 100 per cent embryonic death, whereas daily supplementation with 20 to 45 μg permitted survival of the young, some of which were abnormal (32).

Similar types of anomalies have been reported in the young when a pantothenic acid antimetabolite, pantoyltaurine, was incorporated in the diet of pregnant rats by Zunin and Borrone (33). On the other hand, a chronic or acute pantothenic acid deficiency of pregnant rats in the Long-Evans strain resulted in young exhibiting these anomalies, as well as defects of the interventricular septum of the heart and the aortic arch pattern, hydronephrosis and hydroureter, clubfoot, tail defects, cleft palate, and epidermal defects (34). The acute deficiency was induced by the addition of the antimetabolite, omega-methyl-pantothenate, to the vitamin-deficient regimen for only 2 or 3 days during the second week of pregnancy.

Pteroylglutamic Acid and Vitamin B_{12}

In the studies reported on the teratogenic effects of pteroylglutamic acid (PGA) deficiency in rats, the diets employed have sometimes been sufficiently low in vitamin B_{12} that this deficiency was or may have been a complicating factor. Such studies will be referred to as a combined deficiency of both vitamins.

Richardson and Hogan (35) demonstrated that maintenance of rats for long periods on a purified casein-containing diet low in both PGA and B_{12} resulted in hydrocephalus in nearly 2 per cent of the offspring. Similar findings were observed in an unrelated colony of rats by Richardson and DeMottier (36). The addition of low levels of the crude PGA-antagonist, x-methyl-PGA, to this diet or to a soybean oil meal diet resulted in 25 per cent hydrocephalus by the fourth generation (37). Supplementation of the casein-containing diet with folic acid (PGA) or addition of vitamin B_{12} to the soybean oil meal diet prevented the occurrence of this anomaly (38–40). The critical period for production of hydrocephalus under these experimental conditions was shown to be the second week of pregnancy, in that beginning B_{12} supplementation by the seventh day prevented the abnormality, whereas a delay in supplementation to the fourteenth day was ineffective. Occasional anomalies observed, in addition to hydrocephalus in the young, were spina bifida, cranium bifidum, anophthalmia or microphthalmia, cleft palate, short mandible, and edema (39).

Hydrocephalus appeared to be of the communicating type as there was a marked increase in cerebrospinal fluid pressure as determined by cisternal puncture (39). Morphological study of the ventricular system in hydrocephalic brains (41) showed a transient closure of the cerebral aqueduct between the sixteenth and eighteenth days of gestation. Later the aqueduct was occluded in some cases but more often was merely reduced in size. The lateral ventricles, intraventricular foramen, and third ventricle were greatly distended. The cerebral cortex was decreased in thickness and in some areas

the ependyma was missing (42). Non-hydrocephalic young which had litter-mate hydrocephalic siblings were not able to learn a maze as quickly as con-trol rats, and this impairment of learning capacity was not altered by feeding high levels of PGA after birth (43). The non-hydrocephalic young showed little difference in electroencephalographs in comparison with controls, whereas the hydrocephalic young exhibited a marked reduction in wave frequency, regularity, and amplitude (44). Bruemmer *et al.* (45) demon-strated that there was no cellular difference in deoxyribonucleic acid (DNA) concentration in the brains of hydrocephalic (B_{12}-deficient) and control young. The average cell size was smaller in hydrocephalic young, and, there-fore, the total DNA content of the brain was higher than normal. Histologic studies of B_{12}-deficient rats at birth (46) and during the first 4 weeks of life (47) showed a general retardation of organ maturity accompanied by reduced organ weight, reduced glycogen, and increased fat content of heart, liver, and kidney.

An extensive series of studies on the effects of PGA deficiency during em-bryonic development in the rat have been carried out by Nelson, Evans, and co-workers. The vitamin was shown to be essential for normal reproduction when a purified diet containing 1 per cent succinylsulfathiazole (SST) to pre-vent intestinal vitamin synthesis was used (48, 49). Addition of 0.5 per cent of the crude PGA-antagonist, x-methyl-PGA, accentuated the deficiency and resulted in 100 per cent embryonic mortality at term when the diet was given only during pregnancy (49) or was begun as late as the ninth day of gestation (50). However, when the diet was started on the tenth or eleventh days, 95–100 per cent of the young exhibited marked abnormalities, namely, marked edema and anemia, multiple skeletal defects, retarded development of the viscera (especially kidney and lungs), and lenticular cataract. When the PGA-deficient diet was delayed until the thirteenth day of gestation, only 35 per cent of the young showed mild abnormalities, principally edema and visceral retardation. Normal young resulted when the PGA-deficient regimen was not started until day 15 of gestation. Vitamin supplementation prevented all effects of the PGA deficiency, and embryonic development was normal.

The effects of a transitory PGA deficiency were then studied by giving the PGA-deficient regimen for one, two, or three days, followed by the PGA-sup-plemented diet for the remainder of the gestation period (51). A deficiency of only 48 hours during the second week of pregnancy, the critical period of differentiation and organogenesis, resulted in 70–100 per cent abnormal young or fetal death. In contrast, a 24-hr period of deficiency during this week had practically no effect on embryonic development nor did a 72-hr deficiency period during the first week of pregnancy. The earlier stages of embryonic development were more severely affected by the same length of deficiency than the later phases. The embryonic age most susceptible to PGA deficiency was found to be the ninth day (52).

The types and incidence of abnormalities resulting from a transitory PGA deficiency varied with the time of instituting the deficiency and with its duration or severity. Severe cerebral and eye anomalies were observed only when the PGA-deficient regimen was given from days 7 to 9, whereas cardiovascular defects occurred when the diet was instituted on the seventh, eighth, ninth, or tenth days. The transitory deficiency period from days 9 to 11 resulted in abnormal young with multiple skeletal anomalies including cleft palate and harelip, vascular and urogenital defects, umbilical and diaphragmatic herniations, and diverse retinal and lenticular anomalies of the eye. Some of these anomalies did not occur with a PGA-deficiency period beginning on the tenth day, and only skeletal anomalies were observed with transitory PGA deficiency one day later (days 11 to 14).

The observed cardiovascular defects included interventricular septal defects, persistent truncus arteriosus, and various anomalies of the aortic arch, such as double or right aortic arch, absence of the ductus arteriosus, and aberrant origins of the subclavian and pulmonary arteries (53). Many additional cardiovascular defects have been observed by histological study (54–56). Abnormalities of the urinary tract observed were renal and ureteric hypoplasia, renal ectopia, hydronephrosis and hydroureter, and apparent absence of a kidney. Histological study of these abnormalities revealed additional anomalies and demonstrated that the majority of anomalies appeared to result from retardation or arrest of normal developmental processes (57). Renal ectopia could be explained by retardation of development of both the urinary tract and the vertebral column (58). Skeletal abnormalities were observable grossly in practically all PGA-deficient groups and staining with alizarin red revealed many additional defects (59). The histological study of abnormal bones suggested that although the early stages of mesenchymal condensation were apparently normal, the later stages of chondrogenesis and the resorption of cartilage which proceeds endochondral ossification were adversely affected by PGA deficiency instituted as late as the eleventh day of pregnancy. The early or first effects of a transitory PGA deficiency on the rat embryo were shown (60) to be delay in closure of neural tube and reduced size of the embryo. Three days after the transitory deficiency was instituted, several organs were misshaped and malpositioned—possibly as a result of the observed sparsity of mesenchyme. In addition, a partial metaphase block was observed in conjunction with reduced cytoplasmic basophilia and with reduced number and regularity of staining of ribosomes (61). The observed delays in structural development were accompanied by delays in biochemical development as indicated by asynchronous appearance of multiple molecular forms of enzymes (62).

SUMMARY

The teratogenic effects of five vitamin deficiencies in the rat have been extensively studied. With chronic deficiencies, a typical pattern or syndrome of abnormalities can frequently be produced, e.g., the skeletal malformations in riboflavin-deficient young or the usually observed anomalies of the soft tissues and eyes in vitamin A-deficient embryos. When acute deficiencies are used or vitamin supplementation is given during the second week of pregnancy, diverse anomalies of many types result in the young, particularly in PGA or vitamin E deficiency. The use of antimetabolites has facilitated the production of acute deficiencies and made it possible to study the effects of a transitory deficiency on embryonic development. With antimetabolites, it can be demonstrated that the teratogenic effects vary with the time of instituting the deficiency and with its duration of severity. It can also be demonstrated that the period of sensitivity to the teratogenic dietary deficiencies is usually the critical period of differentiation and organogenesis.

REFERENCES

1. CHAMBERLAIN, J. G. AND NELSON, M. M. Congenital abnormalities in the rat resulting from single injections of 6-aminonicotinamide during pregnancy. *J. Expt'l Zool.* **153:** 285, 1963.
2. PINSKY, L. AND FRASER, F. C. Production of skeletal malformations in the offspring of pregnant mice treated with 6-aminonicotinamide. *Biol. Neonat.* **1:** 106, 1959.
3. KALTER, H. AND WARKANY, J. Experimental production of congenital malformations in mammals by metabolic procedure. *Physiol. Rev.* **39:** 69, 1959.
4. WARKANY, J. AND SCHRAFFENBERGER, E. Congenital malformations induced in rats by maternal nutritional deficiency. VI. The preventive factor. *J. Nutrition* **27:** 477, 1944.
5. WARKANY, J. AND ROTH, C. B. Congenital malformations induced in rats by maternal vitamin A deficiency. II. Effect of varying the preparatory diet upon the yield of abnormal young. *J. Nutrition* **35:** 1, 1948.
6. WILSON, J. G., ROTH, C. B., AND WARKANY, J. An analysis of the syndrome of malformations induced by maternal vitamin A deficiency. Effects of restoration of vitamin A at various times during gestation. *Am. J. Anat.* **92:** 189, 1953.
7. WARKANY, J. AND SCHRAFFENBERGER, E. Congenital malformations of the eyes induced in rats by maternal vitamin A deficiency. *Proc. Soc. Exp. Biol. Med.* **57:** 49, 1944.
8. WARKANY, J. AND SCHRAFFENBERGER, E. Congenital malformations induced in rats by maternal vitamin A deficiency. I. Defects of the eye. *Archiv. Ophthal.* **35:** 150, 1946.
9. WILSON, J. G. AND WARKANY, J. Aortic arch and cardiac anomalies in the offspring of vitamin A deficient rats. *Amer. J. Anat.* **85:** 113, 1949.
10. WILSON, J. G. AND WARKANY, J. Cardiac and aortic arch anomalies in the offspring of vitamin A deficient rats correlated with similar human anomalies. *Pediatrics* **5:** 708, 1950.
11. WILSON, J. G. AND WARKANY, J. Epithelial keratinization as evidence of fetal vitamin A deficiency. *Proc. Soc. Exp. Biol. Med.* **64:** 419, 1947.
12. WILSON, J. G. AND WARKANY, J. Malformations in the genito-urinary tract induced by maternal vitamin A deficiency in the rat. *Amer. J. Anat.* **83:** 357, 1948.
13. EVANS, H. M., BURR, G. O., AND ALTHAUSEN, T. L. The antisterility vitamin fat soluble E. *Memoirs Univ. Calif.* **8:** 1, 1927.

14. MASON, K. E. A hemorrhagic state in the vitamin E-deficient fetus of the rat. Essays in Biology in Honor of Herbert M. Evans. Univ. of Calif. Press, Berkeley, pp. 401, 1943.

15. THOMAS, B. H. AND CHENG, D. W. Congenital abnormalities associated with vitamin E malnutrition. *Proc. Iowa Acad. Sci.* **59**: 218, 1952.

16. CHENG, D. W. AND THOMAS, B. H. Relation of time of therapy to teratogeny in maternal avitaminosis E. *Proc. Iowa Acad. Sci.* **60**: 290, 1953.

17. CHENG, D. W., CHANG, L. F., AND BAIRNSON, T. A. Gross observations on developing abnormal embryos induced by maternal vitamin E deficiency. *Anat. Record* **129**: 167, 1957.

18. WARKANY, J. AND NELSON, R. C. Appearance of skeletal abnormalities in the offspring of rats reared on a deficient diet. *Science* **92**: 383, 1940.

19. WARKANY, J. AND NELSON, R. C. Skeletal abnormalities in the offspring of rats reared on deficient diets. *Anat. Record* **79**: 83, 1941.

20. WARKANY, J., NELSON, R. C., AND SCHRAFFENBERGER, E. Congenital malformations induced in rats by maternal nutritional deficiency. II. Use of varied diets and of different strains of rats. *Amer. J. Dis. Child.* **64**: 860, 1942.

21. WARKANY, J. AND SCHRAFFENBERGER, E. Congenital malformations induced in rats by maternal nutritional deficiency. V. Effects of a purified diet lacking riboflavin. *Proc. Soc. Exp. Biol. Med.* **54**: 92, 1943.

22. WARKANY, J. AND NELSON, R. C. Skeletal abnormalities induced in rats by maternal nutritional deficiency. Histologic studies. *Arch. Pathology* **34**: 375, 1942.

23. WARKANY, J., NELSON, R. C., AND SCHRAFFENBERGER, E. Congenital malformations induced in rats by maternal nutritional deficiency. III. The malformations of the extremities. *J. Bone Joint Surgery* **25**: 261, 1943.

24. WARKANY, J., NELSON, R. C., AND SCHRAFFENBERGER, E. Congenital malformations induced in rats by maternal nutritional deficiency. IV. Cleft palate. *Am. J. Dis. Child.* **65**: 882, 1943.

25. WARKANY, J. AND DEUSCHLE, F. M. Congenital malformations induced in rats by maternal riboflavin deficiency: Dento-facial changes. *J. Am. Dental Assoc.* **51**: 139, 1955.

26. GIROUD, A. AND BOISSELOT, J. Repercussions de l'avitaminose B₂ sur l'embryon du rat. *Archives Franc. de Pediat.* **4**: 317. 1947.

27. NELSON, M. M., BAIRD, C. D. C., WRIGHT, H. V., AND EVANS, H. M. Multiple congenital abnormalities in the rat resulting from riboflavin deficiency induced by the antimetabolite galactoflavin. *J. Nutrition* **58**: 125, 1956.

28. KALTER, H. AND WARKANY, J. Congenital malformations in inbred strains of mice induced by riboflavin-deficient, galactoflavin-containing diets. *J. Exper. Zool.* **136**: 531, 1957.

29. BOISSELOT, J. Malformations congénitales provoquées chez le rat par une insuffisance en acide pantothenique du régime maternal. *Compt. Rend. Soc. Biol.* **142**: 928, 1948.

30. BOISSELOT, J. Malformations foetales par insuffisance en acide pantothenique. *Archives Franc. de Pediat.* **6**: 1, 1949.

31. BOISSELOT, J. Rôle tératogens de la deficience en acide pantothenique chez le rat. *Ann. de Med.* **52**: 225, 1951.

32. BOISSELOT, J. Contribution à l'étude des besoins en acide pantothenique de l'embryon et du foetus. Résultat expérimentaux obtenus chez le rat. *Archiv. des Sciences Physiologiques* **9**: 145, 1955.

33. ZUNIN, C. AND BORRONE, C. Embriopatie da carenza di acido pantotenico. Effetto della pantoiltaurina, antivitamina dell'acido pantotenico. *Acta Vitaminal.* **8**: 263, 1954.

34. NELSON, M. M., WRIGHT, H. V., BAIRD, C. D. C., AND EVANS, H. M. Teratogenic effects of pantothenic acid deficiency in the rat. *J. Nutrition* **62**: 395, 1957.

35. RICHARDSON, L. R. AND HOGAN, A. G. Diet of mother and hydrocephalus in infant rats. *J. Nutrition* **32**: 459, 1946.

36. RICHARDSON, L. R. AND DeMOTTIER, J. Inadequate maternal nutrition and hydrocephalus in infant rats. *Science* **106**: 644, 1947.

37. HOGAN, A. G., O'DELL, B. L., AND WHITLEY, J. R. Maternal nutrition and hydrocephalus in newborn rats. *Proc. Soc. Exp. Biol. Med.* **74**: 293, 1950.

38. O'DELL, B. L. WHITLEY, J. R., AND HOGAN, A. G. Relation of folic acid and vitamin A to incidence of hydrocephalus in infant rats. *Proc. Soc. Exp. Biol. Med.* **69**: 272, 1948.

39. O'DELL, B. L., WHITLEY, J. R., AND HOGAN, A. G. Vitamin B₁₂, a factor in prevention of hydrocephalus in infant rats. *Proc. Soc. Exp. Biol. Med.* **76**: 349, 1951.

40. RICHARDSON, L. R. Nutritional hydrocephalus in infant rats. *Proc. Soc. Exp. Biol. Med.* **76**: 142, 1951.

41. OVERHOLSER, M. D., WHITLEY, J. R., O'DELL, B. L., AND HOGAN, A. G. The ventricular system in hydrocephalic rat brains produced by a deficiency of vitamin B₁₂ or of folic acid in the maternal diet. *Anat. Record* **120**: 917, 1954.

42. NEWBERNE, P. M. AND O'DELL, B. L. Histopathology of hydrocephalus resulting from a deficiency of vitamin B₁₂. *Proc. Soc. Exp. Biol. Med.* **97**: 62, 1958.

43. WHITLEY, J. R., O'DELL, B. L., AND HOGAN, A. G. Effect of diet on maze learning in second-generation rats. Folic acid deficiency. *J. Nutrition* **45**: 153, 1951.

44. OVERHOLSER, M. D., WHITLEY, J. R., O'DELL, B. L., AND HOGAN, A. G. Comparison of electroencephalographs of young rats from dams on synthetic and on normal diets. *Science* **111**: 65, 1950.

45. BRUEMMER, J. H., O'DELL, B. L., AND HOGAN, A. G. Maternal vitamin B₁₂ deficiency and nucleic acid content of tissues from infant rats. *Proc. Soc. Exp. Biol. Med.* **88**: 463, 1955.

46. JONES, C. C., BROWN, S. O., RICHARDSON, L. R., AND SINCLAIR, J. G. Tissue abnormalities in newborn rats from vitamin B₁₂ deficient mothers. *Proc. Soc. Exp. Biol. Med.* **90**: 135, 1955.

47. JOHNSON, E. M. A histologic study of postnatal vitamin B₁₂ deficiency in the rat. *Amer. J. Path.* **44**: 73, 1964.

48. NELSON, M. M. AND EVANS, H. M. Reproduction in the rat on purified diets containing succinylsulfathiazole. *Proc. Soc. Exp. Biol. Med.* **66**: 289, 1947.

49. NELSON, M. M. AND EVANS, H. M. Pteroylglutamic acid and reproduction in the rat. *J. Nutrition* **38**: 11, 1949.

50. NELSON, M. M., ASLING, C. W., AND EVANS, H. M. Production of multiple congenital abnormalities in young by maternal pteroylglutamic acid deficiency during gestation. *J. Nutrition* **48**: 61, 1952.

51. NELSON, M. M., WRIGHT, H. V., ASLING, C. W., AND EVANS, H. M. Multiple congenital abnormalities resulting from transitory deficiency of pteroylglutamic acid during gestation in the rat. *J. Nutrition* **56**: 349, 1955.

52. NELSON, M. M., WRIGHT, H. V., BAIRD, C. D. C., AND EVANS, H. M. Effects of 36-hour period of pteroylglutamic acid deficiency on fetal development in the rat. *Proc. Soc. Exp. Biol. Med.* **92**: 554, 1956.

53. BAIRD, C. D., NELSON, M. M., MONIE, I. W., AND EVANS, H. M. Congenital cardiovascular anomalies induced by pteroylglutamic acid deficiency during gestation in the rat. *Circulation Research* **2**: 544, 1954.

54. MONIE, I. W., BAIRD, C. D., NELSON, M. M., AND EVANS, H. M. Cardiovascular malformations in rat fetuses from mothers suffering transitory pteroylglutamic acid deficiency during pregnancy. *Anat. Record* **121**: 409, 1955.

55. MONIE, I. W., KAROVOCHOS, V., AND NELSON, M. M. Abnormal pulmonary veins in rat fetuses as a result of transitory maternal pteroylglutamic acid deficiency. *Anat. Record* **124**: 424, 1956.

56. MONIE, I. W., NELSON, M. M., AND EVANS, H. M. Persistent right umbilical vein as a result of vitamin deficiency during gestation. *Circulation Research* **5**: 187, 1957.

57. MONIE, I. W., NELSON, M. M., AND EVANS, H. M. Abnormalities of the urinary system

of rat embryos resulting from maternal pteroylglutamic acid deficiency. *Anat. Record* **120**: 119, 1954.

58. MONIE, I. W., NELSON, M. M., AND EVANS, H. M. Abnormalities of the urinary system of rat embryos resulting from transitory deficiency of pteroylglutamic acid during gestation. *Anat. Record* **127**: 711, 1957.

59. ASLING, C. W., NELSON, M. M., WRIGHT, H. V., AND EVANS, H. M. Congenital skeletal abnormalities in fetal rats resulting from maternal pteroylglutamic acid deficiency during gestation. *Anat. Record* **121**: 755, 1955.

60. JOHNSON, E. M., NELSON, M. M., AND MONIE, I. W. Effects of transitory pteroylglutamic acid (PGA) deficiency on embryonic and placental development in the rat. *Anat. Record* **146**: 215, 1963.

61. JOHNSON, E. M. Effects of maternal folic acid deficiency on cytologic phenomena in the rat embryo. *Anat. Record* **149**: 49, 1964.

62. JOHNSON, E. M. Electrophoretic analysis of abnormal development. *Proc. Soc. Exp. Biol. Med.*, in press.

ELECTROPHORETIC AND HISTOCHEMICAL ANALYSES OF EMBRYONIC TISSUES; BREEDING AND MAINTENANCE OF RATS

This outline is annotated relevant to points of special interest or difficulty which are not generally discussed. In addition, citations to the literature have been added both for areas of rather broad background information and also for specific details and protocols.

I. Animal Care
    A. Routine Feeding and Maintenance
        1. Stock Diets (1)

All but the rather large establishments will probably find it unfeasible to produce their own stock diet of natural foodstuffs and will rely on an outside source. The manufacturing firms will supply an analysis of the diet on request, but the investigator must realize that raw materials undergo seasonal variations and will cause unannounced changes in the nutritive value of the ration.

As a further note of caution, dietary requirements are well known to change markedly during gestation. In addition, a state of nutrition which permits reproduction does not necessarily provide for maximum growth and survival of the young, or for optimal animal condition or *in vivo* storage as the young reach experimental age. The condition of the animal as it enters an experiment is of paramount importance even if it is not a nutritional experiment, as subtle areas of possible synergism between the marginal nutritional states and various drugs and stresses are not well understood.

2. Sanitation

   Methods of achieving a desirable level of sanitation will vary greatly, depending on the layout of the physical plant and the animal cages (1). Fresh bedding, food, and clean cages are mandatory. The entrance must be at one side of the animal room, and the refuse must leave from the other side. These two sides should be separated by benches employed as a work area.

3. Temperature and Humidity Control (2)

4. Light and Noise Sequences (2)

B. Normal Pregnancy

1. Age: Weight Ratio of Female

   Females should be chosen with regard to both their weight and their age. Either criterion alone does not provide sufficient control of animal condition as either may affect results (see chap. 3). The maternal weight change during pregnancy, similar to weight gain of individuals prior to pregnancy, must be judged on the basis of prior experience with a particular colony of rats, and not on the basis of a different strain of the same species.

2. Breeding Performance Record of Male

   Male rats vary considerably in their ability to reproduce and, therefore, it is of value to keep accurate records of the reproductive performance. A card is kept (Fig. 1) for each male rat. This card bears his date of birth, number, and earmarkings on a preprinted outline. The number of each female with which he has been placed is entered along with the date. The next day, an unstained smear of vaginal contents is examined with a microscope. If a vaginal plug and/or spermatozoa are found, checks are put in the "P" and/or "S" columns. If cornified cells are also found, it means that the preceding day the female was proestrous ("0" in local laboratory jargon) and the remaining two columns are marked. Positive marks: *1* in the first three or all four columns indicates breedings; *2* in only the first two columns indicates that a proestrous female was provided but the male had failed to function, and if the performance is repeated with a subsequent proestrous female, he should be discarded; if no *3* appears in any column, this indicates that the female was not proestrous and a technician did not correctly diagnose the estrous state (3).

3. Diagnosis and Timing of Pregnancy

   Pregnancy is considered to be at day 0 and hour 0 at 10 A.M. at the day of finding sperm in a smear of vaginal contents. For the remainder of a normal pregnancy, the vaginal smear each day will contain only epithelial cells and leukocytes, except on day 13, when maternal erythrocytes (the placental sign) will be encountered. The presence of red cells prior or subsequent to this time is taken to be indicative of one or more resorbing sites.

4. Care at Littering

At littering, the mother will usually eat severely malformed young in addition to her normal ingestion of the placenta and attached membranes. It is therefore necessary to deliver young by

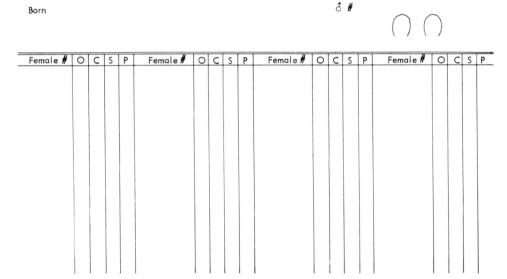

Fig. 1.—Breeding performance record of male. This card is a modification of the type employed in the laboratory of the late M. M. Nelson of the Institute of Experimental Biology and Department of Anatomy, University of California, Berkeley.

caesarian section at day 20 to keep an accurate ratio of malformed:normal:resorbed young.

II. Experimental Protocol

   A. Composition (1)

   B. Feeding and Food-Intake Regulation

It is of particular importance in nutritional experiments to test for any effects due to inanition by means of pair-feeding and/or food restriction. In such techniques the experimental rat is carefully matched by an experimental control animal of the same age and weight but one day less advanced in pregnancy. The weight of the diet eaten by the experimental rat on one day is the weight of the diet offered to the control rat on the next day. Measurement of food intake is attained by weighing the diet cup and its contents each day. The diet cup should be designed to prevent spillage or contamination. Because rats are coprophagous they must be kept on wire screening.

C. Records of Female Condition
    See I, B, and Figure 2.

III. Removal and Study of Embryos
  A. Harvesting Embryos
    1. Excision of Uterine Horns
      The uterine horns should be carefully cleaned and the corpora lutea and resorbing sites counted. Even at day 20, sites of very

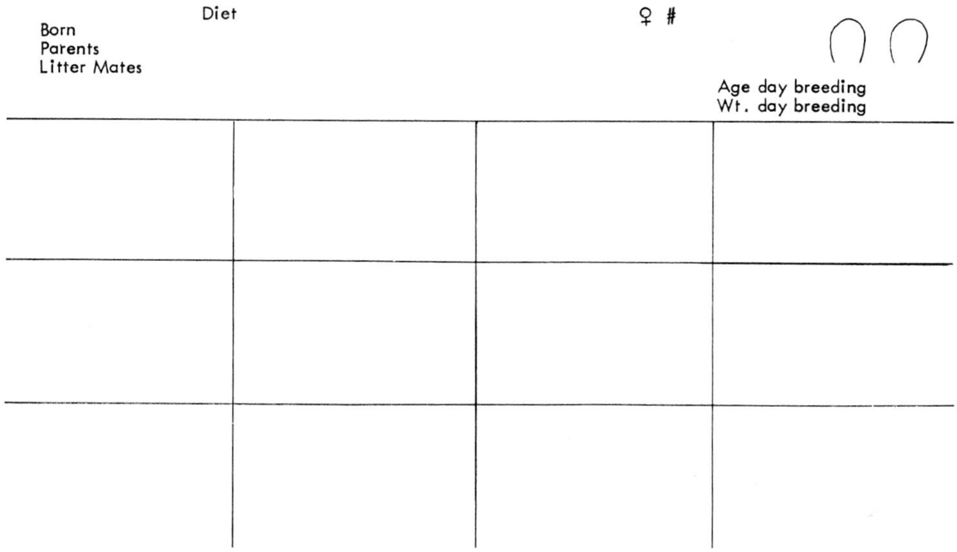

Fig. 2.—Female record card. This record is a modification of the type used in the laboratory of the late M. M. Nelson.

early resorption can be detected as small swellings at the antimesometrial side of the uterus and by remnants of increased vascular supply from ovarian/uterine vessels into localized areas of uterus.

    2. Extirpation of Embryo and Separation of Yolk Sac and Amnion
      Obtaining embryos toward the end of the embryonic period, i.e., days 12, 13, and 14, is no problem; however, at days 10 and 11 the embryos are somewhat more difficult to see and handle. As it is this period of early organogenesis that is of major interest in this discussion, we must devise a method to harvest such embryos.

    With the aid of a dissecting microscope, the antimesometrial myometrium is incised lengthwise, the injury-potential retracts the musculature and exposes the decidua capsularis. The decidua capsularis is then incised, and the yolk sac exposed. The yolk sac is white and shows an indistinct organization of vessels and blood islands in contrast to the granular and reddish decidua. The yolk sac is in turn incised with one tine of a scissors and a rubber bulb-equipped pipette inserted into the site as gentle suction is applied. The embryo and some adhering membranes can then be transferred in the pipette to a suitable fluid and cleaned. Maternal blood is washed free by dilution, the yolk sac is cut away, and the amnion pulled free. Even though the amnion is quite thin, it is easily removed, as it sticks tenaciously to any metal instrument it contacts. If one plans to do enzyme studies on homogenates of isolated embryos, it is important to be meticulous in removing every trace of the yolk sac, as it is enzymatically reactive.

B. Electrophoretic Techniques
  1. Homogenization of Tissues (6)
  2. Preparation of Gel and Insertion of Sample (7, 8)

      A measured sample of the homogenized tissue is placed on a precut piece of filter paper (usually Whatman No. 1) or is pipetted directly into starch grains packed into a slit in the gel. Gels can be made with almost any buffer, just as long as its composition does not interfere with subsequent analyses. However, the pH must be sufficiently removed from the isoelectric point of the protein under investigation to permit migration. In compounding a buffer one must strike a compromise between a concentration low enough to preclude salting out but high enough to carry a current.

  3. Regulation of Electrophoretic Field Strength and Quantification (8)

      Even with a regulated direct-current power supply, it is necessary to have an independent voltmeter/ammeter to monitor the field strength in the individual gels. Obviously, with several gels in parallel, such an independent meter is required to indicate which gels require additional filter paper bridges (8) to lower their resistance and thereby raise their field strength to the desired level. Toward the end of an electrophoretic separation, the gels heat up slightly; and because the buffer is continually being decomposed, the current will need to be adjusted frequently to maintain constant field strength in the gels. Even slight heating can denature the proteins you wish to examine, so if a higher voltage is required, place the apparatus in a cold environment.

There are several methods available for quantification of enzyme activity as represented in zymograms. The oldest is an elution technique whereby the gel is cut in half lengthwise and one-half stained to find the precise location of the protein bands and the other half frozen. The frozen half can then be cut into segments on the basis of the measurements from the stained gel and the enzymes eluted and analyzed by standard biochemical methods. Another method is to cut the stained bands from the starch and place them in solution or suspension for densiometry. As the amount of dye deposited within the gel under standard conditions of time, temperature, etc., is quite reproducible, this is an acceptable method though fraught with several causes of variation even within one day's analysis. A simpler method is to clear the gel in glycerine and pass it between a photocell and a light source, the former being equipped with an integrator read-out source to draw a curve of optical density equivalent to dye deposit. The area under peaks in the curve is then an index of enzyme activity insofar as this is relative to dye deposition (8).

4. Staining of Gels (9, 10)
5. Electrophoretic Analysis of Normal and Abnormal Embryos (11)
C. Electron Microscopic Technique
   1. Preparation of Tissues and Staining

   The method of obtaining embryos is similar to that described but osmium or permanganate fixation is used and it is not necessarily desirable to separate the embryo from its membranes.

   When viewing a section of an embryo with the aid of the electron microscope, it is necessary to cut an adjacent section of that same embryo a little thicker and stain it for light microscopy. This thicker section is viewed through a light microscope placed at the side of the electron microscope and is used for positive orientation to the various embryonic structures observed in the severely restricted field of view in the electron microscope.

   2. Analysis of Normal and Abnormal Embryos at the Ultrastructural Level (12, 13)

REFERENCES

1. *The rat in laboratory investigation,* edited by E. J. FARRIS AND J. Q. GRIFFITH. Philadelphia: Lippincott, 1949.
2. *Methods in animal experimentation,* edited by W. L. GAY. New York: Academic Press, in press.
3. LONG, J. A. AND EVANS, H. M. The estrous cycle in the rat and its associated phenomena. *Mem. Univ. Calif.* **6:** 1, 1922.
4. JOHNSON, E. M., NELSON, M. M., AND MONIE, I. W. Effects of transitory pteroylglutamic

acid (PGA) deficiency on embryonic and placental development in the rat. *Anat. Record* **146**: 215, 1963.

5. JOHNSON, E. M. A histologic study of postnatal vitamin B_{12} deficiency in the rat. *Amer. J. Path.* **44**: 73, 1964.
6. GREGG, J. H., HACKNEY, A. L., AND KRIVANEK, J. O. Nitrogen metabolism of the slime mold *Dictyostelium discoideum* during growth and morphogenesis. *Biol. Bull.,* **107**: 226, 1954.
7. SMITHIES, O. Zone electrophoresis in starch gels; group variations in the serum proteins of normal human adults. *Biochem. J.* **61**: 629, 1955.
8. JOHNSON, E. M. Quantification of enzyme activity subsequent to zone electrophoresis in starch gel. *Analytical Biochem.* **8**: 210–216, 1964.
9. HUNTER, R. L. AND MARKERT, C. D. Histochemical demonstration of enzymes separated by zone electrophoresis. *Science* **125**: 1294, 1957.
10. MARKERT, C. L. AND MØLLER, F. Multiple forms of enzymes: tissue, ontogenetic, and species specific patterns. *Proc. Nat'l Acad. Sci. U.S.* **45**: 753, 1959.
11. JOHNSON, E. M. Electrophoretic analysis of abnormal development. (Submitted to *Proc. Soc. Expt'l Biol. Med.*)
12. JOHNSON, E. M. Effects of maternal folic acid deficiency on cytologic phenomena in the rat embryo. *Anat. Record,* **149**: 49–55, 1964.
13. JOHNSON, E. M. AND CALLAHAN, W. P. Ultrastructure of the early rat embryo, manuscript in preparation.

CHAPTER 6

TERATOLOGICAL STUDIES WITH EXPLANTED CHICK EMBRYOS

NORMAN W. KLEIN

It is unfortunate that the publication explosion allows little journal space for the scientific community to evaluate negative results. Although there are a few exceptions (1–3), the frequently discussed but seldom published "Journal of Negative Results" would probably carry no less than five additional reports indicating that thalidomide does not have specific effects on the development of the chick embryo. These observations serve only to confirm the obvious fact that at the organismic level, the embryological development of man and the chick are different. However, as one progresses from the intact animal through the tissue, cellular, and molecular levels of organization, these sharp differences fade. It is in evaluating the mechanisms underlying the specificity of drug action at these three levels of organization that the chick embryo can serve as a model system for experimental teratology.

The first studies with explanted chick embryos were reported by Waddington in 1932 (4). Although there have been several modifications (5–7) in the procedure, there are two basic steps. The embryo in combination with extraembryonic membrane tissue is first removed from the egg and this is followed by placement onto a semisolid nutrient medium for cultivation. The number of explants that can be prepared in a day indicates that these steps as stated are oversimplified and directs attention to one of the major limitations of the explantation technique. Approximately 30 explants per day can be prepared by a well-trained technician. In comparison, it should not be difficult to handle at least 500 eggs by common egg-injection procedures.

In addition to the amount of time involved, the poor survival and growth of embryos in culture relative to the *in ovo* condition might also limit the applicability of explants to a particular problem in teratology. It has only been during the last 2 years that we have been able to formulate conditions

Dr. Klein was Research Assistant, Institute of Cellular Biology, The University of Connecticut, Storrs, Connecticut when this paper was presented. He is now Assistant Professor, Department of Biology, Marquette University, Milwaukee, Wisconsin.

This work was supported by research grants to Dr. Heinz Herrmann from the Institute of Neurological Diseases and Blindness, U.S. Public Health Service (B-2238), and The Association for the Aid of Crippled Children.

which permit embryos to survive for 72 hours *in vitro*. Previously, chick embryo explants survived approximately 24 hours. After 72 hours of cultivation, under optimum conditions the growth of explants in terms of protein-nitrogen content is approximately one-third that of embryos incubated for comparable periods *in ovo*.

In spite of these limitations, chick embryo explants offer several advantages over other methods frequently employed with the chick embryo. Such factors as differences in development at the time of oviposition (8) lead to variations in physiological age after a particular period of incubation. By microscopically examining embryos prior to explantation it is possible to select a uniform group of organisms for experimentation. This selectivity reduces variations in response and greatly facilitates the opportunity to conduct quantitative evaluations of the response. Employing the injection procedure frequently leads to problems concerning variations in the site of administration, diffusibility of the substance injected, and limitation of the quantities which can be administered. Explants obviously minimize these problems and provide the additional opportunity to transfer embryos to different media after various periods of exposure.

NUTRITIONAL CONTROL

It is possible to culture embryos under conditions of precise nutritional control. Precise implies that the endogeneous supply of nutrients is limited to the extent that the embryo is dependent on the medium. This is an important consideration in dealing with the influence of various substances on the embryo because a large supply of endogenous nutrients could lead to unknown dilution factors and limit the uptake of a compound.

The studies on nutritional control were combined with investigations on the ability of embryos to utilize a chemically defined medium (9). Attempts were first made to culture embryos with large extra-embryonic membranes (10) on a defined medium and to demonstrate growth inhibition by omitting single essential amino acids. These studies failed because considerable amounts of yolk were transferred with the embryo from the egg to the culture. This yolk could not be removed from the extra-embryonic membrane by simple rinsing. Thus, it became necessary to limit the size of the membrane.

When this was done, it was possible to demonstrate that both protein and DNA synthesis were inhibited by the omission of single essential amino acids from the chemically defined medium (Table 1). Furthermore, the extent of inhibition resulting from either a leucine or lysine deficiency approximated the extent of inhibition observed when all the amino acids were omitted from the medium (basal). The growth of embryos on the whole-egg medium

(50 per cent fresh egg and 50 per cent agar salts) was greater than that observed on the chemically defined medium, but survival was inversely related to growth. The morphological development of the embryos, as judged by somite number, was independent of growth.

The data are only presented for embryos evaluated after 48 hours of cultivation. The analysis after 24 hours provided similar differences but smaller in magnitude. However a comparison of the growth parameters of embryos cultured on the basal medium for 24 hours with the protein nitrogen and DNA of embryos at zero time (11–13 somites) indicated that nitrogenous precursors were still available from the endogenous and residual yolk pools. Thus, during the first 24 hours of cultivation the protein and DNA doubled in the complete absence of exogenous nitrogen. The limited growth between 24 and 48 hours of cultivation on either the basal or amino acid-deficient media demonstrated that the endogenous pools of nutrients become exhausted during the first 24 hours of cultivation.

TABLE 1

EMBRYOS EXPLANTED FOR 48 HOURS ON VARIOUS MEDIA[a]

Media	Protein Nitrogen (μg/embryo)	DNA (μg/embryo)	Somites	Survival (%)
Whole-egg homogenate	22	13	38	24
Basal (amino acid-free)	10	7	34	67
Chemically defined	17	10	36	50
Chemically defined minus leucine	11	7	35	53
Chemically defined minus lysine	10	7	34	55
Chemically defined minus proline and aspartic acid	15	10	36	69

[a] Data from Klein *et al.* (9).

GASEOUS ENVIRONMENT

Although chick embryo explants have been employed successfully as a model system to investigate many aspects of embryology including teratology (Table 2), poor growth and survival have limited the scope of analysis and have caused difficulties in quantitative evaluation. In an attempt to overcome these difficulties, embryos were cultured in high levels of oxygen (7). The first studies involving either 100 per cent O_2 or 95 per cent O_2 + 5 per cent CO_2 clearly indicated that oxygen would greatly enhance growth and prolong survival of embryos. An intensive series of investigations was then performed in which the amounts of oxygen, carbon dioxide, and air were varied during different intervals of the cultivation periods. Results from this series indicated that when chick embryos of 11 to 13 somites are cultured on the whole-egg homogenate medium, gassing with 75 per cent air + 25 per cent O_2 for the first 24 hours of cultivation followed by 95 per cent O_2 + 5

TABLE 2

EFFECT OF VARIOUS COMPOUNDS ON THE EXPLANTED CHICK EMBRYO

Abbr.: ABB, *Arch. Biochem. Biophys.*; BB, *Biol. Bull.*; DB, *Develop. Biol.*; ECR, *Exptl. Cell Res.*; G, *Growth*; JEEM, *J. Embryol. Exptl. Morphol.*; JEZ, *J. Exptl. Zool.*; JGP, *J. Gen. Physiol.*; S, *Science*.

Compound	Pre-incubation or Stage of Explantation	Results	Reference
Acetylpyridine	11–13 somites	Differential inhibition of protein and DNA synthesis. TPN generating system evaluated.	Newburgh *et al.*, ABB **97**, 94.
Actinomycin D	11–13 somites	Inhibits growth and development in posterior region of body axis.	Klein and Pierro, S **142**, 967.
Amino acid analogues (valine, leucine, phenylalanine, and methionine)	Primitive streak to 8 somites	Inhibition related to specific analogue. Somite segmentation blocked; shortening of body, and abnormalities of brain, neural tube, and somites.	Rothfels, JEZ **125**, 17. Herrmann, JEEM **1**, 291. Herrmann *et al.*, JEZ **128**, 359.
Amino acid and purine analogues	Primitive streak or head process	Various abnormalities and growth inhibition.	Waddington and Perry, JEEM **6**, 365.
Aminopterin	21 hr.	Brain, heart channels, heart, and somites inhibited. Protection with DNA and RNA plus thymidine.	O'Dell and McKenzie, JEEM **11**, 185.
Azaguanine	Primitive streak to early somite	Somite number reduced, embryo shorter, and vesicles on sides of posterior axis.	Frair and Woodside, G **20**, 9.
Antimycin A and sodium fluoride	18–22 hr.	Nutrition, site of administration, and developmental age influences site of action. Sodium fluoride sensitive areas related to prospective heart regions.	Duffey and Ebert, JEEM **5**, 324. McKenzie and Ebert, JEEM **8**, 314.
Bromoallylglycine	32 hr.	Calcium-activated ATPase activity of somite mesoderm reduced. Leucine reverses. Catheptic activity of somite mesoderm increased.	Deuchar, JEEM **8**, 259. Deuchar, DB **2**, 129.
Bromoallylglycine and fluorophenylalanine	11–13 somites	Inhibition of protein synthesis and catheptic activity.	Hahn and Herrmann, DB **5**, 309.
Bromoallylglycine	11–13 somites	Reduction in protein content of somites and primitive knot. Reduces uptake of tracer glycine into primitive knot.	Schultz and Herrmann, JEEM **6**, 262.
Bromoallylglycine	4–7 somites	Pycnosis in mesoderm of posterior half of embryo.	Schultz, ECR **17**, 353.
Butyrobetaine	Primitive streak to 4 somites	Inhibits development and causes degeneration. Protection with carnitine.	Ito and Fraenkel, JGP **41**, 279.
Chloroacetophenone	Primitive streak and head process	Inhibition of brain at early stage. Loss of induction capacity of Hensen's node.	Lakshmi, JEEM **10**, 373. Lakshmi, JEEM **10**, 383.
Deoxyguanylate	11–13 somites	Inhibits growth and deoxycytidylate deaminase.	Roth *et al.*, S **142**, 1473.
Fluorophenylalanine	11–13 somites	Protein increase inhibited more than incorporation of tracer glycine.	Herrmann and Marchok, DB **7**, 207.
Insulin	24 hr.	Brain and neural tube inhibited at low concentrations. Somites and and heart at higher concentrations.	Barron and McKenzie, JEEM **10**, 88.
Lithium chloride	Primitive streak to somites	Cyclopia.	Rogers, DB **8**, 129.
Mercaptoethanol, dithidiglycol, and lipoic acid	Primitive streak to 4 somites	Neural tube closure inhibited. Incorporation of nucleic acid precursors evaluated.	Rohl and Brachet, DB **4**, 549. Pohl and Quertier, JEEM **11**, 293.

TABLE 2—*Continued*

Compound	Pre-incubation or Stage of Explanation	Results	Reference
Metabolic inhibitors (iodoacetate, fluoride, citrate, malonate, cyanide, and azide	Primitive streak to 8 somites	Differential degeneration of heart and nervous system.	Spratt, BB **99**, 120
Nitrogen mustard derivatives	Primitive streak	Degeneration of neural tube and somite cells.	Jurand, JEEM **8**, 60. Jurand, JEEM **9**, 492.
Puromycin	11–13 somites	Inhibits growth and development in posterior region of body axis.	Klein and Pierro, unpublished.
Triethanomelamine	22–23 hr.	Retardation of development. Somites sensitive.	Jurand, JEEM **6**, 357.
Trypan blue	Primitive streak and medullary plate	Malformation of somites, heart, and notochord. Short body axis and large blisters.	Mulherkar, JEEM **8**, 1.
Trypan blue	24–72 hr.	Various abnormalities related to stage of explantation.	Stephan and Sutter, JEEM **9**, 410.

per cent CO_2 for the 24- to 72-hr period provided optimum conditions in the gaseous phase for growth, development, and survival.

Comparisons of the results obtained with oxygen and air indicate the magnitude of the oxygen response (Table 3). In view of the poor survival of embryos cultured in air for 48 hours, no attempt was made to maintain these embryos for 72 hours. However, after 48 hours, the growth parameters of embryos cultured in oxygen were more than double the values obtained from embryos provided with air. With air, only 24 per cent of the embryos survived 48 hours of culture, but oxygen gave 100 per cent survival after

TABLE 3[a]

EFFECT OF OXYGEN ON THE EXPLANTED CHICK EMBRYO WITH
WHOLE-EGG HOMOGENATE MEDIUM

	O TIME	GAS[b]	Hours *in Vitro*		
			24 hr	48 hr	72 hr
Protein nitrogen (μg/embryo)	4	A	16	22	
		O	20	52	77
DNA (μg/embryo)	3	A	8	13	
		O	12	27	39
Somites	11–13	A	29	38	
		O	28	42	
Survival (per cent)		A	100	24	
		O	100	100	86
Stage (23)		O		19	22–23

[a] Data from Klein (7, 9).

[b] A = air; O = 24 hours of 75 per cent air + 25 per cent O_2, 48 and 72 hours of 95 per cent O_2 + 5 per cent CO_2.

48 hours and 86 per cent after 72 hours. Few embryos survived through the fourth day, and these exhibited degenerative changes.

Various problems related to the oxygen response have been investigated. The use of 95 per cent $O_2 + 5$ per cent CO_2, or 100 per cent O_2 from the start of the cultivation period leads not only to reduced growth as compared to 75 per cent air $+ 25$ per cent O_2 but also to an inhibition in the development of the circulatory system. By comparing the inhibition of blood-pigment formation in embryos of different initial somite numbers, this response was related to the developmental stage. Embryos explanted with more than 13 pairs of somites did not exhibit a reduction in blood pigment.

Embryos developing *in ovo* were exposed to oxygen in the manner found optimum for cultured embryos. Oxygen did not enhance the growth or development of embryos *in ovo* but did increase the variability of the protein nitrogen and DNA content of the embryos. The oxygen response of cultured embryos appears, therefore, to be related to the *in vitro* condition. The embryo depends on the circulation of the extra-embryonic membrane (yolk sac) for gaseous exchange at this stage. This membrane develops extensively *in ovo* but not under culture conditions. This suggests that oxygen might simply compensate for the limited circulatory systems available for respiratory functions.

The major emphasis in our investigations has been directed toward problems of nutrition. The growth of embryos on a chemically defined medium is not enhanced by oxygen. The addition of horse serum or embryo extract to the defined medium does not alter the response. During the first 24 hours of cultivation, the circulatory system becomes functional. By transferring embryos from the whole-egg medium to the defined medium after 24 hours of cultivation, it was possible to evaluate the role of the circulatory system in the utilization of free amino acids. The values presented (Table 4) indicate that the growth of embryos on a chemically defined medium was not enhanced by a pre-cultivation period on the egg medium. However, the data do illustrate that the maintenance of nutritional control during such a transfer experiment is possible and that this manipulation is not deleterious (note whole-egg controls).

The differences in the concentrations of nitrogenous precursors between the defined and whole-egg media (0.169 g/100 ml of amino acids versus 7 g/100 ml of protein) suggested a possible explanation for the inadequacy of the defined medium. Higher levels of amino acids were found toxic, but dilution of the egg medium indicated that at equivalent levels of nitrogen, the egg-nutrient medium supported growth to the same extent as the defined medium. Since the hypothesis that high concentrations of amino acids were essential for the rapid growth of embryos and that protein only served to provide high levels of amino acids in a form that was compatible with survival was workable, attempts were made to culture embryos on various sources of protein. The results indicated that a purified protein such as ovalbumin could support the growth and development of explanted embryos when combined with low molecular weight components of egg yolk and albumen (11).

In concluding this section it might be helpful to present some values for protein nitrogen which will place the growth of explants under two conditions of nutrition in proper perspective to the growth of embryos *in ovo* (Table 5).

Although these values indicate that it has not been possible to duplicate the *in ovo* growth with cultured embryos, the opportunity to study development under a variety of growth conditions provides a unique tool for experimentation. For example, in the near future we plan to investigate the relationship of growth to teratological specificity.

TABLE 4

PROTEIN NITROGEN AND DNA CONTENT OF CHICK
EMBRYOS TRANSFERRED TO VARIOUS MEDIA
AFTER 24 HOURS OF CULTIVATION ON WHOLE-
EGG HOMOGENATE MEDIUM[a]

Media	Protein Nitrogen (μg/embryo)	DNA (μg/embryo)
48 hours:		
Whole-egg homogenate.....	46	25
Chemically defined........	23	16
Basal (amino acid free).....	21	13
72 hours:		
Whole-egg homogenate.....	74	39
Chemically defined........	25	16
Basal (amino acid free).....	18	12

[a] Data from Klein (7).

TABLE 5[a]

PROTEIN NITROGEN CONTENT OF CHICK EMBRYOS CULTURED
in Vitro AND INCUBATED *in Ovo*[b]

Culture and Incubation Conditions	24 hr (μg/embryo)	48 hr (μg/embryo)	72 hr (μg/embryo)
in ovo	29	106	292
whole-egg medium *in vitro*	20	52	77
chemically defined medium *in vitro*	12	17	

[a] Data from Klein (7, 9).
[b] Zero time = 40 hours of incubation or 11–13 somites.

Actinomycin D

Actinomycin D is one of the most popular compounds in biological research today. Its popularity is derived from the observations that this antibiotic binds specifically with deoxyguanosine of the DNA molecule (12, 13). This binding blocks the synthesis of DNA-dependent RNA. In a series of two papers Pierro reported (14, 15) that the injection of small amounts of antinomycin into the yolk sac of chick embryos during the second day of incubation caused abnormalities associated with the posterior regions of the axial skeleton. Six hours after administration, cell death and degenerative changes could be detected in the unsegmented somite mesoderm and undifferentiated nodal mass. Injection after 32 hours of incubation produced a greater frequency of more severe axial defects (trunklessness) than after 48 hours (rumplessness).

The main reason for employing chick embryo explants to evaluate the action of actinomycin D was to express in quantitative biochemical terms the degree of teratogenic specificity (16). Chick embryos of 11 to 13 somites were cultured for 48 hours on the whole-egg medium containing various amounts of this compound. Gross observations suggested that the development of the tail region was inhibited in proportion to the levels of actinomycin D in the nutrient medium. For the purpose of quantitation, embryos were separated into an anterior portion consisting of head, heart, and axial tissue to the twelfth somite and a posterior fraction consisting of axial tissue posterior to the twelfth somite. Dissection, therefore, separated the tissue which appeared to be the main site of drug inhibition (unsegmented somite mesoderm and trunk-tail node) from the tissue which appeared resistant. As was expected from morphological observations, the protein nitrogen, DNA, and RNA contents of the posterior fractions were lower in the groups receiving actinomycin D than in the controls (Table 6). However, simple morphological inspection could not serve to evaluate the response of the anterior fraction. The quantitative evaluations indicated that the anterior fractions as well as the membranes were not appreciably inhibited by quantities of actinomycin D which blocked the development of the posterior axial tissue. Such quantitative determinations of the regional specificity of drug action should clearly indicate the advantages of chick embryo explants in this type of investigation.

Two possible explanations were considered in the initial interpretation of the actinomycin response. First, these observations could simply indicate that the compound was not uniformly distributed due to either differences in cell permeability or physiological direction. The second, based on current biochemical dogma, suggested that in the transformation of cells from unsegmented to segmented somite mesoderm a stable form of messenger RNA is synthesized. However, this latter observation was recently negated by the observation that puromycin duplicated the regional specificity of actinomycin inhibition in the explanted chick embryo.

Although both actinomycin and puromycin ultimately cause an inhibition of protein synthesis, the sites at which this inhibition occurs are different. Actinomycin presumably inhibits the synthesis of messenger RNA while puromycin competes with aminoacyl-soluble RNA for ribosomal sites (17). Thus, in order to investigate the mechanisms underlying the action of these drugs, one is confronted with a basic problem in embryology, namely, factors which control the rates of protein synthesis. The plea presented by Windle (18) for investigations into "basic parameters of embryogenesis" appears most appropriate.

TABLE 6

EFFECT OF ACTINOMYCIN D ON EXPLANTED CHICK EMBRYOS[a]

Dosage (μg/ml = μg/embryo)	Protein Nitrogen (μg/embryo)	DNA (μg/embryo)	RNA (μg/embryo)
Posterior Fraction:			
0	9	7	9
0.125	7	6	6
0.25	5	5	3
0.50	2	4	< 2
Anterior Fraction:			
0	39	21	38
0.125	40	23	39
0.25	40	22	37
0.50	34	18	33
Membrane:			
0	79	26	56
0.125	87	27	54
0.25	80	24	50
0.50	70	18	37

[a] Data from Klein and Pierro (16).

Further studies on actinomycin and puromycin must first involve an evaluation of distribution. These studies might be followed by precursor incorporation studies or growth versus specificity investigations. For these experiments, chick embryo explants are particularly well suited.

However, in pursuing the mechanism of drug specificity, the trail must eventually lead to an analysis of the unique properties of a particular group of cells. A small embryo frequently presents a formidable problem associated with providing sufficient quantities of material from a particular cell population for analysis. In addition, it is difficult to control and evaluate such problems as precursor pools in an intact organism. For these reasons, we are presently attempting to establish cell cultures derived from regions of drug resistance and drug sensitivity. Although teratology in cell cultures is another facet of this problem, the studies of Hsu and Somers (19) and Kit *et al.* (20) with bromodeoxyuridine and the work of Barban and Schulze (21) and Barban (22) with deoxyglucose should be mentioned. In these studies cell cultures were used to investigate the mechanisms of drug action and the mechanisms of drug resistance.

REFERENCES

1. VERRETT, J. AND McLAUGHLIN, J. Use of the chick embryo technique in the evaluation of the toxicity of drugs. *Fed Proc.* **22:** 188, 1963.
2. SHORB, M. S., SMITH, C., VASAITIS, V., LUND, P. G., AND POLLARD, W. Effect of thalidomide treatment of hens on embryonic development and fertility. *Proc. Soc. Exptl. Biol. Med.* **113:** 619, 1963.
3. SALZGEBER, B. AND SALAUN, J. Limb malformations obtained in the chick embryo after thalidomide treatment. *Excerpta Medica* **3:** 1185, 1963.
4. WADDINGTON, C. H. Experiments on the development of chick and duck embryos cultivated *in vitro*. *Phil. Trans. Roy. Soc. London Ser. B* **221:** 179, 1932.
5. SPRATT, N. T. Development *in vitro* of the early chick blastoderm explanted on yolk and albumen extract saline-agar substrate. *J. Exptl. Zool.* **106:** 345, 1947.
6. NEW, D. A. T. A new technique for the cultivation of the chick embryo *in vitro*. *J. Embryol. Exptl. Morphol.* **3:** 326, 1955.
7. KLEIN, N. W., McCONNELL, E., AND RIQUIER, D. J. Enhanced growth and survival of explanted chick embryos cultured under high levels of oxygen. *Develop. Biol.* **10:** 17, 1964.
8. TAYLOR, L. W. Development of the unincubated chick embryo in relation to hatchability of the egg. *Proc. Seventh World's Poultry Congress.* 1939, p. 188.
9. KLEIN, N. W., McCONNELL, E., AND BUCKINGHAM, B. J. Growth of explanted chick embryos on a chemically defined medium and effects of specific amino acid deficiencies. *Develop. Biol.* **5:** 296, 1962.
10. BRITT, L. G. AND HERRMANN, H. Protein accumulation in early chick embryos grown under different conditions of explantation. *J. Embryol. Exptl. Morphol.* **7:** 66, 1959.
11. KLEIN, N. W. Growth of chick embryo explants on various sources of protein, in preparation.
12. KIRK, J. M. The mode of action of actinomycin D. *Biochim. Biophys. Acta* **42:** 167, 1960.
13. KERSTEN, W. Interaction of actinomycin C with constituents of nucleic acids. *Biochim. Biophys. Acta* **47:** 610, 1961.
14. PIERRO, L. J. Teratogenic action of actinomycin D in the embryonic chick. *J. Exptl. Zool.* **147:** 203, 1961.
15. PIERRO, L. J. Teratogenic action of actinomycin D in the embryonic chick. II. Early development. *J. Exptl. Zool.* **148:** 241, 1961.
16. KLEIN, N. W. AND PIERRO, L. J. Actinomycin D: Specific inhibitory effects on the explanted chick embryo. *Science* **142:** 967, 1963.
17. NATHANS, D. AND NEIDLE, A. Structural requirements for puromycin inhibition of protein synthesis. *Nature* **197:** 1076, 1963.
18. WINDLE, W. F. *Report of the conference on prenatal effects of drugs.* Chicago: Commission on Drug Safety, p. 46.
19. HSU, T. C. AND SOMERS, C. E. Properties of L cells resistant to 5-bromodeoxyuridine. *Exptl. Cell. Res.* **26:** 404, 1962.
20. KIT, S., DUBBS, D. R., PIEKARSKI, L. J., AND HSU, T. C. Deletion of thymidine kinase from L cells resistant to bromodeoxyuridine. *Exptl. Cell Res.* **31:** 297, 1963.
21. BARBAN, S. AND SCHULZE, H. O. The effects of 2-deoxyglucose on the growth and metabolism of cultured human cells. *J. Biol. Chem.* **236:** 1887, 1961.

22. BARBAN, S. Studies on the mechanism of resistance to 2-deoxy-D-glucose in mammalian cell cultures. *J. Biol. Chem.* **237**: 291, 1962.

23. HAMBURGER, V. AND HAMILTON, H. L. C. A series of normal stages in the development of the chick embryo. *J. Morphol.* **88**: 49, 1951.

CHICK EMBRYO EXPLANTATION TECHNIQUE; YOLK SAC PERFUSION TECHNIQUE

CHICK EMBRYO EXPLANTATION

Fertile eggs are incubated for approximately 40 hours. The precise period of pre-incubation varies with the season and holding temperature. The objective is to obtain the largest possible number of embryos of the desired developmental age.

The eggs are wiped with 70 per cent ethanol before the contents are released. The shell is cracked with a suitable instrument (e.g., Bard-Parker handle), and the contents are placed in an evaporating dish (e.g., Corning 3180; top diameter, 125 mm; height, 65 mm) containing about 20 ml of chick Ringer's solution. Care must be taken to avoid rupture of the vitelline membrane. The yolks and albumen from about eight eggs are placed in the dish. Further manipulations are greatly facilitated by filling the evaporating dish to capacity.

The vitelline membrane is held securely with forceps in the region of the area vitellina, and a circular incision is made within the area vitellina proximal to the embryo. If the incision is made on the distal side of the area vitellina, it is difficult to remove the vitelline membrane from over the embryo and extra-embryonic membrane. It is important to make the circular incision complete. The cut edge, including the extra-embryonic and vitelline membranes, is grasped with forceps and draw into a spoon. It is important to include the vitelline membrane with the embryo and extra-embryonic membrane because the extra-embryonic membrane is easily torn. A convenient spoon can be made by flattening the end and bending up the edges of a "Scoopit" stainless steel spatula.

The embryo, including vitelline and extra-embryonic membranes, is placed in a Petri dish containing chick Ringer's salt solution. All embryos contained in the evaporating dish are excised before proceeding to the next step. Forceps and gentle agitation are used to remove the vitelline membrane from over the embryo. Embryos are selected with the aid of a dissecting microscope and transmitted light (25×). We select embryos of 11 to 13 somites.

The extra-embryonic membrane peripheral to the sinus terminalis is next trimmed off and discarded. Care is taken not to cut into the sinus terminalis. The embryo, including the remaining portion of the extra-embryonic membrane, is drawn up into a large-bore bulb pipette (diameter at orifice about 5 mm) and transferred through three separate dishes containing chick Ringer's salt solution. The embryo and extra-embryonic membrane are rinsed by repeated pipette aspiration in each of the three dishes.

Before the embryo is placed on the nutrient medium, a few drops of clean chick Ringer's are placed on the medium. The embryo is placed ventral surface down onto the medium. The edge of the membrane curls in the dorsal direction making it simple to determine the proper surface to place down. The embryo is flattened on the surface of the medium with a few drops of Ringer's and the excess chick Ringer's is drawn off with a small-orifice bulb pipette.

The culture dishes consist of a Petri dish (Corning 3160, top diameter, 100 mm; depth of lower dish, 10 mm) and a watch glass (diameter, 50 mm) supported by a cotton ring. About 3 ml of Ringer's is added to the cotton ring for moisture. The watch glass contains 1 ml of nutrient medium.

The gassing chambers are made from Turtox plastic laboratory dishes, size F, 10 inches in diameter and 3.5 inches in depth. The lid of the plastic dish is fitted with three stoppers containing glass tubes. The center stopper and tube serve as the gas inlet and two stopper–tubes placed peripherally on the lid serve as gas outlets. The gas is directed to the bottom of the plastic dish where it is first bubbled through a water reservoir prior to being exposed to the embryos. The gas is removed from the incubator or hot room (37.5° C) through rubber tubing connections from the two outlet tubes to an aspirator. A flow rate of 1 ft^3/hr for each plastic dish is maintained. Sixteen chick embryo explants contained in individual Petri dish cultures can be placed in each plastic dish. Four empty Petri dishes are used as a base in the plastic dish, and the lid of the plastic dish is held secure with tape. A gas mixture of 75 per cent air + 25 per cent O_2 (volume per cent) is passed through the plastic dishes for the 0- to 24-hr culture period and then a gas mixture of 95 per cent O_2 + 5 per cent CO_2 is employed.

The whole-egg homogenate medium is made by combining the contents of fresh egg in equal volume with an agar Ringer's solution. The contents of a fresh egg are placed in a beaker and thoroughly mixed. The agar solution consists of 1.5 g agar + 100 ml of chick Ringer's. The agar solution is autoclaved and brought to a temperature between 40° and 50° C before combining with the egg. An equal volume of the fresh egg and agar is mixed together and 1 ml aliquots are dispensed into the watch glasses. Before the medium hardens, pieces of lens paper of 1.5–2.0 cm^2 are placed on the medium. The medium is prepared the day before explantation and stored in a refrigerator.

For transfer studies, the lens tissue squares with adhering embryo and extra-embryonic membrane are removed from the medium and placed in Ringer's. The embryo and membrane are separated from the lens paper by gentle agitation and the embryo with membrane can then be handled with a large-bore pipette.

After the desired period of cultivation, the embryo and extra-embryonic membrane are usually placed on black rubber stoppers for dissection. A standard dissection consists of separating the embryo from extra-embryonic membrane at the edge of the somites.

The chick Ringer's solution contains in grams per liter: NaCl, 7; KCl, 0.42; and $CaCl_2$, 0.24. Antibiotics are added to the nutrient medium and chick Ringer's solution used for rinsing in grams per liter as follows: dihydrostreptomycin sulfate, 0.660; penicillin G potassium, 0.0060.

SELECTED BIBLIOGRAPHY

BRITT, L. G. AND HERRMANN, H. Protein accumulation in early chick embryos grown under different conditions of explantation. *J. Embryol. Exptl. Morphol.* **7**: 66, 1959.

HAYASHI, Y. AND HERMANN, H. Growth and glycine incorporation in chick embryo explants. *Develop. Biol.* **1**: 437, 1959.

KLEIN, N. W., McCONNELL, E., AND BUCKINGHAM, B. J. Growth of explanted chick embryos on a chemically defined medium and effects of specific amino acid deficiencies. *Develop. Biol.* **5**: 296, 1962.

KLEIN, N. W., McCONNELL, E., AND RIQUIER, D. J. Enhanced growth and survival of explanted chick embryos cultured under high levels of oxygen. *Develop. Biol.,* **10**: 17, 1964.

SPRATT, N. T. Development *in vitro* of the early chick blastoderm explanted on yolk and albumen extract saline-agar substrata. *J. Exptl. Zool.* **106**: 345, 1947.

YOLK SAC PERFUSION

The basic steps involved in preparing an egg for perfusion and replacement are essentially the same. The following description, taken directly from the published papers, will be limited to yolk replacement. It is believed that this latter procedure will be useful to teratological investigations because this method permits the administration of large quantities of materials which will become uniformly distributed in the yolk. It is possible to replace yolk with donor yolk at 4 days of incubation and still hatch chicks. The perfusion procedure is designed to allow continuous or intermittent addition of substances into the yolk sac and is described in the references provided.

Eggs are incubated for 3 days at 37.5° C. The egg is kept on its side to bring the embryo to the top surface. A hole 5 mm in diameter is drilled through the shell, but not through the shell membrane, at about the equator of the egg. The drilling is done from below to prevent damage to the embryo, which has come to lie at the top opposite the hole. A stone 4 mm in diameter powered by a hand-held electric drill is satisfactory.

A cup fashioned from aluminum foil is placed on the indifferent plate of an electrosurgical unit (Blendtome, manufactured by the Birtcher Corporation), and Ringer's is poured into the cup and onto the plate in order to assure a good electrical contact through the bottom third of the egg. The straight, sharp electrode, which is made from a piece of stainless steel rod 2 mm in diameter, projects 22 mm beyond the handle. The electrosurgical unit is set for mixed coagulation (Setting 3) and cutting (Setting 4). The hole is swabbed with 70 per cent ethanol. The egg is moved quickly from a horizontal to a vertical position, placed small end down in the cup, and the electrode is plunged horizontally through the shell membrane and white and into the yolk, as deeply as possible. The current is quickly switched on, the electrode held steady for 6 sec, then withdrawn slowly with a spiraling motion during an additional 6 sec. The egg is immediately placed hole downward again and the shell wiped clean with tissue. If the egg is turned over after the canal is made, the embryo and its embryonic circulation may become detached from the vitelline membrane and die.

A probe is used to clear the canal of coagulated yolk and the egg is supported, hole down, on a ring stand. A hypodermic needle (20 gauge) which is clamped firmly and attached by tubing to a reservoir of salts solution, is inserted about 15 mm into the hole. A pressure of 25–35 cm of water is used to flush out the yolk. After all the original yolk has been flushed out, the egg is placed hole upward. Donor yolk, diluted with a salt solution (50 per cent by volume), is injected through a 16-gauge needle into the yolk cavity. This replaces the salt solution that was used to remove and replace the original yolk. The hole can be sealed with plastic adhesive tape. Eggs are incubated in a horizontal position.

The salt solution suggested for use in the flushing steps and as a diluent for the injected yolk is referred to as "salts 26" and contains in grams per liter: NaCl, 4.46; KCl, 1.70; $MgCl_2$ (6 H_2O), 0.34; $MgSO_4$ (7 H_2O), 0.20; NaH_2PO_4 (H_2O), 0.28; glucose, 5.00; $NaHCO_3$, 1.1; $CaCl_2$, 0.02; erythromycin, 0.095; and dihydrostreptomycin, 0.66.

SELECTED BIBLIOGRAPHY

GRAU, C. R., FRITZ, H. I., WALKER, N. E., AND KLEIN, N. W. Nutrition studies with chick embryos deprived of yolk. *J. Exptl. Zool.* **150:** 185, 1962.

GRAU, C. R., KLEIN, N. W., AND LAU, T. L. Total replacement of the yolk of chick embryos. *J. Embryol. Exptl. Morphol.* **5:** 210, 1957.

GRAU, C. R., WALKER, N. E., FRITZ, H. I., AND PETERS, S. M. Successful development of chick embryos nourished by yolk sac perfusion with calcium-low media. *Nature* **197:** 257, 1963.

KLEIN, N. W., GRAU, C. R., AND GREEN, N. J. Yolk sac perfusion of chick embryos: effects of various media on survival. *Proc. Soc. Exptl. Biol. Med.* **97:** 425, 1958.

WALKER, N. E., AND GRAU, C. R. Growth and development of chick embryos supplied with various concentrations of yolk. *J. Nutri.* **79:** 205, 1963.

CHAPTER 7

FACTORS INFLUENCING TERATOGENIC RESPONSE TO DRUGS

M. LOIS MURPHY

Effects of drugs on proliferating cells in pharmacological research are basic to cancer chemotherapy (1). The drugs presently used in treating cancer act on metabolic pathways identical to pathways of normal cells. Toxicity of varying degrees is expected at therapeutic doses. Intensive study of the mechanism of action of new drugs should improve therapeutic techniques and contribute to the understanding of neoplastic processes.

Biological systems at the molecular, cellular, and tissue levels being studied include a species spectrum of microorganisms, plants, and animals; developing embryos and regenerating structures; neoplasms of various types; organ functions including the marrow, gastrointestinal tract, central and peripheral nervous systems, liver and kidneys; and carcinogenic and mutagenic effects and many others.

Drug-induced malformations have been described in many mammals including man, dogs, swine, rabbits, guinea pigs, hamsters, rats, and mice, and studies have been reported in rhesus monkeys, opossum, and armadillos. The rat, however, has several advantages. Wistar rats are available with a known gestation date. When transported to the laboratory before implantation, there is no interference with the normal process of gestation; implantation of the fertilized ovum occurs on about the eighth day; the period of rapid organogenesis extends from the eighth through about the twelfth day of gestation; the majority litter on the twenty-second day; the usual number of fetuses is 10. By sacrificing on the twenty-first day, the abnormal fetuses can be salvaged from probable destruction by the mother and external malformations can be detected without magnification. Staining and clearing for study of the bones, however, requires removing the skin and viscera.

FACTORS INFLUENCING TERATOGENIC EFFECTS OF A DRUG

There are many variables which can modify a teratogenic effect. The chemical properties and specific biochemical activity at a cellular and molec-

Associate Member, Sloan-Kettering Institute for Cancer Research, 444 E. 68th Street, New York, N.Y.

ular level are discussed elsewhere in this volume. The stage during development of the embryo when the drug is administered to the pregnant animal intercepts a dynamic series of particular biochemical events of qualitative and quantitative nature giving differential susceptibility to injury of molecules, cells, tissues, and organs. There is the influence of the speed and manner of excretion and detoxification of the drug by the pregnant animal, the placenta, and the fetus. The dose of the drug administered and the frequency of administration can be varied. Of importance also is the route of administration: intravenous, intraperitoneal, oral, subcutaneous, or intramuscular. Administration of protecting substances in timed relation to the study drug may modify or cancel a teratogenic effect. The species of animal and the strain may affect results. There are many miscellaneous factors; for example, nutritional and endocrine influences as well as many others still unknown might influence response to a drug.

Selection of chemicals for study.—From those drugs active as growth inhibitors in other systems and particularly from those drugs without cross resistance to known chemotherapeutic agents, study drugs are selected (Tables 1 and 2). They usually are the ones of interest at a given time in screening for new agents with potential activity for cancer chemotherapy. The aim is to find one with new and unique types of activity. Within a family of active chemicals having similar structures and biological effects, there may exist differences in teratogenic effects which may contribute to the eventual identification of mechanism of action of the drug.

Selection of dose.—The initial dose is the acute, single LD_{50} for mature rats. A small number of pregnant rats of 9 to 12 days of gestation receive this dose. If, at 21 days, fetuses are alive and normal in surviving pregnant animals (Plate 1), the drug does not have much interest as a teratogenic agent. In contrast, if all of the fetuses have been resorbed leaving only residual implantation sites in surviving pregnant animals (Plate 2), further studies are pursued.

Small numbers of pregnant rats then receive stepwise lowered doses until one is found which allows a few fetuses of a litter to survive at 21 days, and with teratogenic chemicals, these survivors usually have malformations. If the few survivors are normal, interest in the chemical tends to decrease. However, depending on the magnitude of the range between the dose lethal for the mothers and lethal for the fetuses even without malformations, a selective action against the fetus may be found. Several different methods of treatment have been used to increase the production of malformations in the 21-day fetuses.

Methods of increasing teratogenic effects.—Although the majority of the drug studies carried out have been with a single injection on one day of the gestation, several other dosage schedules were found to produce surviving embryos with abnormalities when the single dose gave complete lethality or a very small number of surviving fetuses or mothers. In addition to the single dose, repeated lower daily doses on several or all days of gestation can be given. If the drug is excreted rapidly, small multiple doses on one day, for instance, 4, 6, 8, or 12 hours apart, are used. A single dose, plus a dose of a protective substance which saves fetuses and sometimes mothers from lethality with some drugs, has allowed abnormal embryos to survive. There can also be a combination of chemicals with additive effects.

Spectrum of sensitivity to injury of the fetus throughout gestation in the newborn and adult rat.—During early development, the fertilized ovum has not been injured by intraperitoneal injections to the pregnant animal of a dose of many chemicals studied. Later on these drugs caused fetal resorption. A certain dose of an active chemical at the time of implantation will cause the animal to lack any sign of pregnancy at sacrifice on the twenty-first day of gestation. Many chemicals at doses causing resorption at the time of implantation, if given on one of the following 3 or 4 days, allow fetuses to survive showing developmental abnormalities at sacrifice on the twenty-first day of gesta-

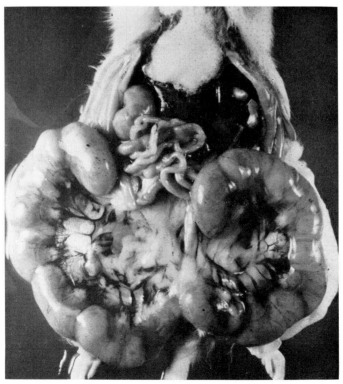

PLATE 1.—A control rat sacrificed on the twenty-first day of gestation shows normal fetuses in the uterus.

tion. In many instances, on these successive, critical days, fetuses tolerate a progressively larger dose. By the fourteenth day of gestation, the fetuses can survive much higher doses. An illustration of studies with azaserine as a single dose at implantation, during rapid organogenesis, at 5 days of age, and in adults is shown in Figure 1 (2). Careful examination of skeletal preparations has revealed abnormalities when none was detected when treated litters were examined for external features on the twenty-first day.

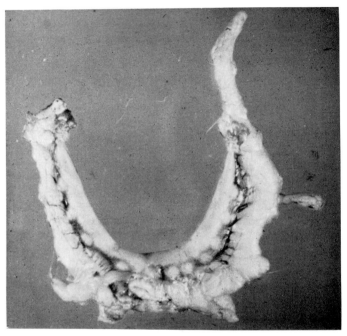

PLATE 2.—The pregnant rat received a dose of a chemical on the eleventh day of gestation which caused, subsequently, complete resorption of the fetuses. The rat was sacrificed on the twenty-first day of gestation, and the uterus in the photograph demonstrates residual implantation sites.

ILLUSTRATIONS OF STUDIES WITH SEVERAL DRUG GROUPS

Alkylating agents.—A series of alkylating agents at varying doses was compared for several effects on the pregnant rat and its fetuses on the twelfth day of gestation (3). The structural formulas are shown in Figure 2. These agents are used in man principally for the lymphomas and there are depressing effects on the normal bone marrow which are reversible. The ratios of several effects as compared to a defined litter LD_{50} are shown in Figures 3 and 4. Tables 1 and 2 give the type and incidence of gross and osseous abnormalities.

Figure 4 shows a set of ratios of several antimetabolites which emphasize the wide range in dosage that produces the defined effects as compared with

M. Lois Murphy   148

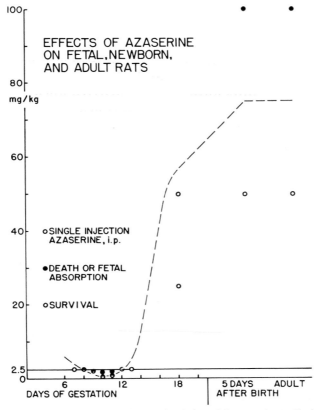

FIG. 1.—This is a comparison of dosages of azaserine injected intraperitoneally into the pregnant animal on various days of the gestation period, in the five-day-old rat and adult rat which allows survival or causes death. Partial litter survival on the ninth through the twelfth days of gestation was accompanied by malformations.

1. bis(β-chlorethyl)methylamine (HN2)

2. p-(N,N-di-2-chlorethyl)amino-phenylbutyric acid (CB1348)

3. Triethylene melamine (TEM)

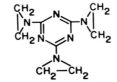

4. Triethylenethiophosphoramide (ThioTEPA)

5. 1,4-Dimethanesulfonyloxybutane (Myleran)

FIG. 2.—Structural formulas of some alkylating agents.

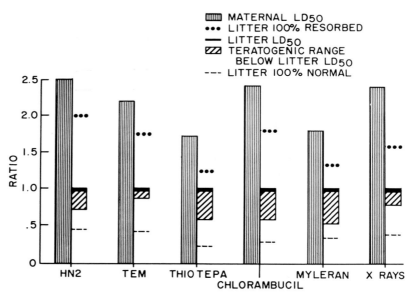

IIIII MATERIAL LD50
••• LITTER 100% RESORBED
— LITTER LD50
▨▨ TERATOGENIC RANGE
     BELOW LITTER LD50
--- LITTER 100% NORMAL

Fig. 3.—Ratio of dosage of alkylating agents and X-rays given pregnant rats on the twelfth day of gestation to produce various effects on mother and fetus to estimated litter LD_{50}.

Litter LD_{50}. A positive litter effect is defined as a dose that is lethal to a majority of the fetuses by the twenty-first day of gestation. At this time, half or more of the implantation sites would contain dead or resorbing fetuses. The litter LD_{50} is the dose calculated to produce a positive litter effect in 50 per cent of the treated mother rats. This value is taken as 1.0 for each compound, and the doses producing other effects are recorded as ratios of the litter LD_{50}.

Maternal LD_{50}. These were estimated for each compound from the toxicity data reported in the literature on non-pregnant rats and were confirmed in our laboratory on small groups of pregnant rats. Pregnancy did not seem to affect the toxicity of the alkylating agents. In determining the maternal LD_{50} for a drug given at the twelfth day of gestation, the observation period was only 10 days, since the rats were sacrificed or littered at this time. On the basis of the 10-day observation period, the estimated maternal LD_{50} ranged from 2.5 times the litter LD_{50} for HN2, chlorambucil, and X-rays to 2.25 for TEM and 1.75 the litter LD_{50} for Thio-TEPA and Myleran. Myleran characteristically causes delayed deaths 14 to 17 days after injection in non-pregnant rats. Therefore, the dose of Myleran lethal to 50 per cent of pregnant rats within 10 days is larger than LD_{50} reported in the literature.

Litter 100 Per Cent Resorbed. The dose giving 100 per cent resorptions was from 1.25 to 2 times the litter LD_{50} for the drugs tested, but always less than maternal LD_{50}.

The Teratogenic Range below Litter LD_{50}. This is charted as the dose range between litter LD_{50} and the lowest dose given in this study at which teratogenic effects were observed.

Litter 100 Per Cent Normal. This is the highest dose administered in these studies that failed to produce teratogenic effects by our methods of observation. The litter 100 per cent normal is about one-third to one-half the litter LD_{50} for the alkylating agents and X-rays.

Drug	Estimated litter LD_{50}
HN2	0.7 mg/kg
TEM	0.55 mg/kg
Thio-TEPA	5.0 mg/kg
Chlorambucil	10.0 mg/kg
Myleran	34.0 mg/kg
X-rays	250 r

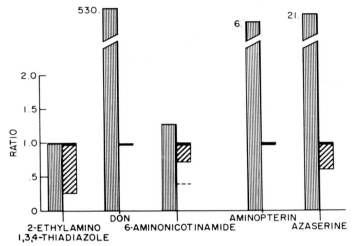

FIG. 4.—Ratio of dosage of several antimetabolites injected intraperitoneally into pregnant rats on the twelfth day of gestation to produce various effects on mother and fetus to estimated litter LD$_{50}$ (see Fig. 3 for an explanation of the symbols and effects described). Brief summaries of pharmacological data on these agents, which have been reported in more detail elsewhere, are given in the text.

2-ETHYLAMINO-1,3,4-THIADIAZOLE is a water-soluble chemical. It inhibits the growth of transplantable tumors, and this can be prevented by the simultaneous administration of nicotinamide. In pregnant rats, a single injection administered on the twelfth day caused gross and skeletal abnormalities of the fetus at doses between 0.25 to 1.0 times the maternal LD$_{50}$ dose (50 to 200 mg/kg). Some of the abnormalities at 100 mg/kg included cleft palate, short lower jaws, short tails, and syndactyly of the front and rear feet. Some of the pregnant animals surviving the maternal LD$_{50}$ contained viable fetuses, indicating that the litter LD$_{50}$ is in the same range as the maternal LD$_{50}$.

DON (6-DIAZO-5-OXO-L-NORLEUCINE) is a water-soluble chemical. It is inhibitory to a wide spectrum of transplantable rat and mouse tumors, and it produces developmental abnormalities in the chick embryo. Some of its effects are prevented by the administration of adenine and 4-amino-5-imidazole carboxamide. In rats, the maternal LD$_{50}$ of DON is 80 mg/kg. When 0.2 mg/kg was given on the twelfth day, the fetuses were all resorbed, but 0.1 mg/kg seemed to have no adverse effect on fetal development and survival. The ratio of litter to maternal LD$_{50}$ for DON is 530:1.

6-AMINONICOTINAMIDE is a powerful antagonist of nicotinamide. It is poorly soluble in water and is prepared as a suspension in 0.5 per cent CMC. The maternal LD$_{50}$ dose for the rat is 8 to 15 mg/kg, being somewhat lower earlier in gestation. At LD$_{50}$ doses, toxic manifestations in the adult include conjunctival hemorrhages (bloody eyes), paralysis of the rear extremities, and weight loss (the paralysis persisted during the life of one animal that survived for 3 months and was then sacrificed). These effects can be prevented by the simultaneous administration of a 100 mg/kg dose of nicotinamide, but administration of the same dose of nicotinamide after paralysis had developed did not correct it. Eight mg/kg doses of 6-aminonicotinamide to the pregnant female on the twelfth day produced maternal weight loss and inhibition of fetal weight, syndactyly, harelip, and cleft palate. Doses of 5 mg/kg produced no fetal resorptions or abnormalities. The ratio of litter to maternal LD$_{50}$ is almost 1:1.

AMINOPTERIN (4-AMINOPTEROYLGLUTAMIC ACID) is a water-soluble antagonist of folic acid. The LD$_{50}$ for the pregnant rat is approximately 3 mg/kg. The litter LD$_{50}$ on the twelfth day of gestation is about 0.4 mg/kg. It is interesting to note that fetuses that survive are usually normal, a situation similar to that seen with DON. The ratio of litter to maternal LD$_{50}$ is from 6 to 8:1.

AZASERINE (O-DIAZO-ACETYL-L-SERINE) is a water-soluble chemical, closely related chemically pharmacologically to DON, but less active by weight. The maternal LD$_{50}$ is 75 to 100 mg/kg as a single dose. The litter LD$_{50}$ was 3.0 mg/kg on the twelfth day, and surviving fetuses usually showed gross and skeletal abnormalities. The latter consisted of fusion of vertebral bodies, small pelvis, and specific deletion of femurs and fibulas. The ratio of fetal to maternal LD$_{50}$ for azaserine is 21:1.

TABLE 1

INCIDENCE OF SELECTED GROSS ABNORMALITIES OF THE RAT FETUS[a]

NUMBER OF FETUSES AND NUMBER OF ABNORMALITIES	HN2		TEM		THIO-TEPA		CHLORAMBUCIL		MYLERAN		X-RAY	
	<LD50[b]	>LD50[c]	<LD50	>LD50	<LD50	>LD50	<LD50	>LD50	<LD50	>LD50	<LD50	>LD50
Number of fetuses examined	66	61	25	35	73	16	59	47	105	17	39	14
Total[d]	127		60		89		106		122		53	
Number with Specific Abnormalities												
Decreased weight and length	66	61	25	35	36	16	59	47	105	17	39	14
Syndactylous forepaws	11	7	24	23	29	13	44	35	88	17	37	14
Short kinky tail	0	6	2	8	3	5	54	38	88	16	30	14
Encephalocele	3	3	6	5	1	3	22	44	0	0	4	14
Syndactylous rear paws	3	2	2	7	0	1	17	38	81	17	34	14
Cleft palate	4	2	0	0	0	0	10	38	12	0	8	14
Edema 4+	0	2	3	9	0	2	7	33	0	1	14	14

[a] Produced by injecting various alkylating agents intraperitoneally into the mother on the twelfth day of gestation.

[b] <LD50: mothers received a dose below the litter LD50, but showing teratogenic activity.

[c] >LD50: mothers received a dose at or above the litter LD50.

[d] Total number of fetuses examined. These fetuses were obtained from litters whose mothers received a dose of a drug that had been demonstrated to produce consistent incidence of teratogenic activity.

TABLE 2

Incidence of Selected Skeletal Abnormalities of the Rat Fetus[a]

Number of Fetuses and Number of Abnormalities	HN2		TEM		Thio-TEPA		Chlorambucil		Myleran		X-Ray	
	<LD_{50}[b]	>LD_{50}[c]	<LD_{50}	>LD_{50}	<LD_{50}	>LD_{50}	<LD_{50}	>LD_{50}	<LD_{50}	>LD_{50}	<LD_{50}	>LD_{50}
Number of fetuses examined	51	57	25	29	20	14	52	30	66	16	37	7
Total[d]	108		54		34		82		82		44	
Number with Specific Abnormalities												
Ribs	18	33	13	3	10	2	28	30	50	16	37	7
Sternum	25	14	8	2	13	2	24	28	35	16	19	7
Zygoma	5	9	11	2	9	2	19	26	19	0	37	7
Occiput	13	16	20	3	8	2	17	30	0	0	23	7
Ulna	6	9	10	4	0	1	20	30	3	11	37	7
Scapula	10	16	4	2	1	0	11	26	30	15	8	7
Radius	3	7	7	2	0	0	11	26	3	12	37	7
Fibula	4	1	8	3	0	0	15	29	11	11	37	7
Tibia	0	0	6	2	1	0	11	23	0	0	37	7
Humerus	0	6	0	0	0	0	7	16	10	8	37	7
Femur	2	0	1	0	0	0	1	6	0	0	37	7

[a] Produced by injecting various alkylating agents intraperitoneally into the mother on the twelfth day of gestation.

[b] <LD_{50}: mothers received a dose below the litter LD_{50}, but showing teratogenic activity.

[c] >LD_{50}: mothers received a dose at or above the litter LD_{50}.

[d] Total number of fetuses examined. These fetuses were obtained from litters whose mothers received a dose of a drug that had been demonstrated to produce consistent incidence of teratogenic activity.

153

the relatively narrow one for the group of alkylating agents (3, 4). Thiersch has studied effects of some of these and other alkylating agents on the rat litter *in utero* (5).

Monie in studying the genitourinary system of rat fetuses whose mothers had received chlorambucil during pregnancy noted that the unilateral or bilateral absence of the kidney was the most frequently encountered abnormality of the system (6). Later on, he studied the 4.5-month-old fetus of a 27-year-old married white woman with Hodgkin's disease who had become pregnant while being treated with chlorambucil. The fetus appeared normal externally but dissection revealed absence of the left kidney and ureter (7).

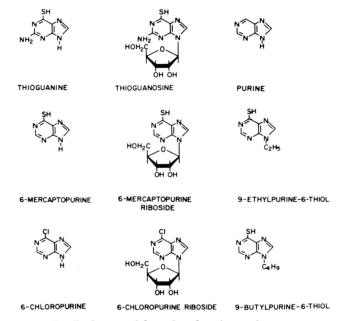

FIG. 5.—Structural formulas of purine analogues.

Sokal reviewed cases of women with lymphomatous diseases who were frequently treated with alkylating or other agents and later became pregnant. Abnormal infants are rarely reported among these women (8). But caution is indicated in that an external normal appearance does not necessarily imply absence of abnormalities of the viscera, skeleton, brain, etc. Miscarriage occurs in many of these patients, and internal examination of the fetus may be entirely lacking.

Purine analogues.—Ten analogues of the physiological purine bases, adenine, guanine, and hypoxanthine, have been studied in 12-day rat fetuses for teratogenic effects (9). The names and structural formulas are shown in Figure 5. These chemicals were developed to interfere with nucleic acid synthesis and inhibit growth of cancer cells. All of them are active in inhibiting animal tumors except chloropurine riboside. In humans, purine analogue toxicity is variable—chloropurine riboside and 9-butyl-6-mercaptopurine having neither activity nor toxicity. Purine, in the few patients treated, produced such a formidable skin rash that trials were terminated and ultimate hematological and intestinal toxicity were not determined. The other seven agents produce gastrointestinal and bone marrow toxicity.

8-Azaguanine was also studied. In humans, it produces mild skin rash and gastrointestinal and bone marrow depression. The main clinical use of purine analogues, however, is the remission produced in acute leukemia. About equally effective in this action, at appropriate doses, are 6-mercaptopurine, 6-thioguanine, and their ribosides (6-chloropurine and 9-ethyl-6-mercaptopurine). Too toxic are 2,6-diaminopurine, 8-azaguanine, and purine. The effects of 6-mercaptopurine on the rat fetus were studied by Thiersch (10).

Figure 6 shows the molar-equivalent dosages covering the teratogenic

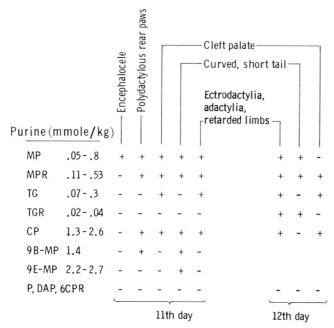

Purine (mmole/kg)		Encephalocele	Polydactylous rear paws	Ectrodactylia, adactylia, retarded limbs	Curved, short tail	Cleft palate		Ectrodactylia, adactylia, retarded limbs	Curved, short tail	Cleft palate
MP	.05 - .8	+	+	+	+	+		+	+	-
MPR	.11 - .53	-	+	+	+	+		+	+	+
TG	.07 - .3	-	-	+	-	+		+	-	+
TGR	.02 - .04	-	-	-	-	-		+	+	-
CP	1.3 - 2.6	-	+	+	+	+		+	-	+
9B-MP	1.4	-	+	-	+	-				
9E-MP	2.2 - 2.7	-	-	-	+	-				
P, DAP, 6CPR		-	-	-	-	-		-	-	-
				11th day					12th day	

Fig. 6.—Purine Analogues. On the left is a list of 10 thio-purines compared for (1) millimoles per kg representing the teratogenic dose range, (2) the abnormalities seen only in fetuses when pregnant animals were treated with chemicals on the eleventh day, and (3) those abnormalities common to fetuses whose mothers were treated on the eleventh or twelfth day of gestation.

MP	6-Mercaptopurine
MPR	6-Mercaptopurine riboside
TG	Thioguanine
TGR	Thioguanosine
CP	6-Chloropurine
9B-MP	9-Butyl-6-mercaptopurine
9E-MP	9-Ethyl-6-mercaptopurine
P	Purine
DAP	2,6-Diaminopurine
CPR	6-Chloropurine riboside

range in laboratory studies of 11- and 12-day pregnant rats. Also abnormalities seen on only the eleventh day and those common to both days are indicated. The analogues without demonstrated teratogenic effects in these rat studies are purine, 2,6-diaminopurine, 8-azaguanine, and 6-chloropurine riboside. The first three give lethal effects and the latter is inactive. On a molecular basis, the most active member of the group that is teratogenic in the rat is thioguanosine and least active is 9-ethyl-6-mercaptopurine.

These purine analogues illustrate that there are subtle intraspecies differ-

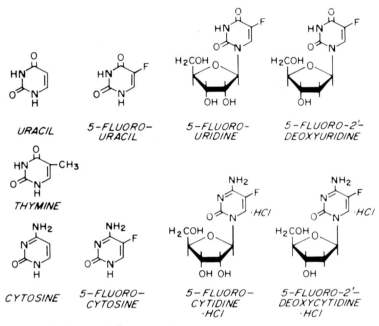

Fig. 7.—Structural formulas of pyrimidines and fluorine analogues.

ences in chemicals that are apparently similar in toxicity and animal tumor-inhibition studies.

Pyrimidine analogues.—Pyrimidine bases (uracil and cytosine) and their nucleosides substituted in the 5 position by fluorine or other halogens (Fig. 7) have been given to pregnant rats by single injections to study teratogenic effects on the fetuses from the ninth through the seventeenth days of gestation (11, 12). These agents are active in inhibiting animal tumors and produce similar patterns of toxicity in animals and man. Two, 5-fluoro-2′-deoxyuridine (FUDR) and 5-fluorouracil (FU), have had widespread testing against cancer in humans.

FUDR interferes with nucleic acid synthesis by interfering with thymidine formation. 5-Fluoro-2′-deoxycytidine (FCDR) is initially deaminated to form

FUDR and then acts in a similar manner. The rat fetuses apparently contain a small amount of deaminase and, therefore, this conversion is slow. Plate 3 shows the comparable doses of the two drugs used and abnormalities seen by single injections on days 9 through 14 of gestation. Figure 8 shows that the single dose lethal for adult rats is greater for FCDR than FUDR but the sensitivity of fetuses is greater for FCDR.

Plate 4 is an example of the skeletal preparations selected as representative

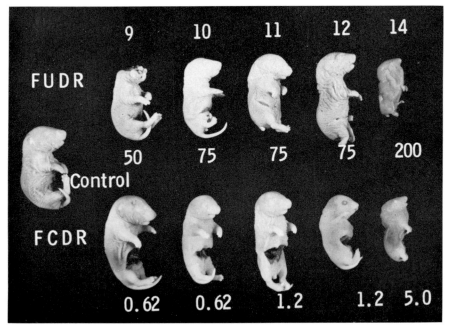

PLATE 3.—5-Fluorodeoxyuridine (FUDR) and 5-fluorodeoxycytidine (FCDR). This is a comparison of the abnormalities caused by various doses of each of the two chemicals shown beneath each fetus when injected into 10 different pregnant rats on one of the days of gestation listed at the top of the photograph. A control fetus is shown at the left. On the ninth day, FUDR at 50 mg/kg to the pregnant animal produced fetuses with severe deformity of the head. FCDR, on the other hand, did not cause any malformations at 0.62 mg/kg and with a slightly higher dose, fetuses were resorbed. With both drugs, some stunting and abnormalities were observed when the pregnant animals were treated on the other days listed.

of each of five different days of gestation from different litters treated with FUDR on different days of gestation. Chloro-, bromo-, and iodo-substituted deoxyuridine inhibit growth by interfering with the utilization of thymidine. They are all teratogenic to rat fetuses at doses to the pregnant rat of 500 to 1000 mg/kg.

Drug-induced abnormalities in two mammalian species.—Diazo-oxo-1-norleucine (DON), a glutamine analogue, has an adult rat LD_{50} of about 80 mg/kg. A dose of 0.2 mg/kg, however, is lethal to 11-day rat fetuses. At

TOXICITY

	FUDR mg/kg		FCDR mg/kg	
40 gm rats	2000	$\dfrac{2\ dead}{2\ inj.}$	2000	$\dfrac{0\ dead}{2\ inj.}$
Fetus Days gestation	M LD_{100} mg/kg pregnant rat			
9	75		1.25	
10	150		1.25	
11	200		5.0	
12	200		5.0	
14	>300		>20.0	

FIG. 8.—Lethality was crudely tested in 40-g rats instead of larger ones, and fluorodeoxyuridine (FUDR) appeared more toxic than fluorodeoxycytidine. The approximated minimal LD_{100} for fetuses for the two compounds on several different days by injection into pregnant rats is shown.

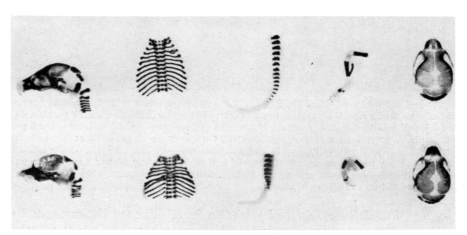

PLATE 4.—5-Fluorodeoxyuridine (FUDR). These are fetuses which have been cleared, stained, and dissected to demonstrate bone abnormalities. The top row are controls. The lower row from left to right are examples of abnormal fetuses whose mothers were treated on day 9, 10, 11, 12, or 17 with doses of 40, 50, 75, 50, or 300 mg/kg or FUDR, respectively, and sacrificed on the twenty-first day. Bone abnormalities are often detected at doses lower than those causing gross malformations. The defects of the fetuses treated on the ninth day are shortness of the maxilla, poorly ossified frontal parietal bones, and hemivertebrae; tenth day, fused ribs; eleventh day, fused or distorted caudal vertebrae; twelfth day, clubbed rear legs and absent tibia; seventeenth day, retarded development of the nasal and frontal bones.

lower doses, fetuses are normal. Protection studies with 100 mg/kg each of adenine and guanine 30 minutes prior to 0.5 mg/kg of DON have allowed rat fetuses to survive and show abnormalities. Plate 5 illustrates the lip and palate abnormalities on each of three days (11).

DON at 0.15 mg/kg on days 20, 21, and 22 of gestation in dogs produced cleft palate, hare lip, abnormal viscera, and deformed legs (13) (Plate 6).

Studies with chlorodeoxyuridine in mice (14) and rats (15) have shown similar abnormalities and protection by administration of thymidine.

SUMMARY

Active chemicals can be studied routinely in mammalian embryos where normal tissue is growing at the accelerated embryonic rate in the same host as adult control tissue. Differential lethality of the two has varied with chemicals from approximately 1:1 to 2000:1 for embryo and adult.

The embryo may survive but show abnormalities of development. Before implantation, the fertilized ovum appears to be relatively resistant or protected. Lethal effects are usual at the time of implantation. The eleventh and twelfth days of gestation in rats are the times when the rat is the most likely to be susceptible to production of drug-induced abnormalities. With differ-

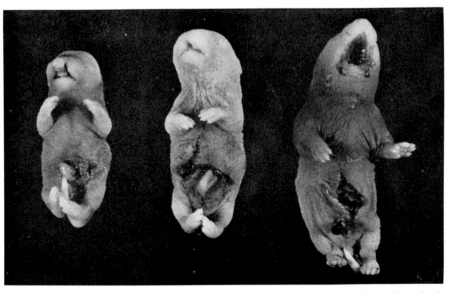

PLATE 5.—DIAZO-OXO-1-NORLEUCINE (DON). The pregnant rats received 100 mg/kg of adenine and guanine one-half hour before receiving 0.5 mg/kg of DON. The fetus on the left, from a litter whose mother was treated on the tenth day of gestation, shows bilateral cleft lip. Cleft palate is also present. The fetus in the center, from a litter whose mother was treated on the eleventh day of gestation, shows a cleft right upper lip. Cleft palate is also present. The fetus on the right is from a litter whose mother was treated on the twelfth day of gestation. It has had the lower jaw removed to demonstrate the cleft palate. The lip is normal.

ent chemical groups, this susceptibility may extend from the eighth through fourteenth and possibly as late as the seventeenth day. Abnormalities involve the skull bones and possibly the brain at day 17. More than one species may show a similar spectrum of drug-induced abnormalities.

Further studies with chemicals of increasingly fundamental action on growth processes should add to understanding of the nature of fertilization, differentiation, implantation, organogenesis, and control mechanisms for normal growth. They may contribute to information on the nature of abnormal growth, including neoplasia. Animal studies do not necessarily predict for the human situation but are not without value. They are the best

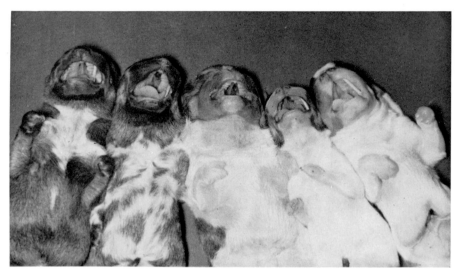

PLATE 6.—These five puppies whelped on the sixtieth day of gestation of a mongrel bitch treated with 0.15 mg/kg of DON three times on days 20, 21, and 22 of gestation. All show cleft palate. They also had abnormal paws, short kinky tails, and two had umbilical hernias (Dr. Mark Friedman).

available at the moment. Human trials will probably always be a reality and must be recognized as such and controlled as carefully as possible. Good biological data are superior to speculation, questions, and laws. The speculation, questions, and laws must be tolerated until the good biological data replace them.

ACKNOWLEDGMENTS

Figures 2, 3, and 4 and Tables 1 and 2 are from MURPHY, M. L., DEL MORO, A., and LACON, C. (see below).

Figure 6, Plates 3, 4, and 5 are from MURPHY, M. L. *Clinical Proceedings of the Children's Hospital of Washington, D.C.*, 1962.

M. Lois Murphy   160

Figure 2 and Plates 4, 5, and 6 are from MURPHY, M. L. *Ciba Foundation Symposium on Congenital Malformations*. London: Churchill, Ltd., 1960.

REFERENCES

1. KARNOFSKY, D. A. Cellular effects of anticancer drugs. *Ann. Rev. Pharmacology* **3**:357, 1963.
2. MURPHY, M. L. AND KARNOFSKY, D. A. Effect of azaserine and other growth-inhibiting agents on fetal development in the rat. *Cancer* **9**: 955, 1956.
3. MURPHY, M. L., DEL MORO, A., AND LACON, C. The comparative effects of five poly-functional alkylating agents on the rat fetus with additional notes on the chick embryo. *Ann. N.Y. Academy Sciences* **68**: 762, 1958.
4. MURPHY, M. L., DAGG, C. P., AND KARNOFSKY, D. A. Comparison of teratogenic chemicals in the rat and chick embryos. *Pediatrics* **19**: 701, 1957.
5. THIERSCH, J. Effects of 2,4,6-triamino-*s*-triazine (TR): 2,4,6 *Tris*-(ethyleneimino)-*s*-triazine (TEM); and N,N', N''-triethylene phosphoramide (TEPA) on rat litter *in utero*. *Proc. Soc. Exp. Biol. Med.* **94**: 36, 1957.
6. MONIE, I. W. Chlorambucil-induced abnormalities of urogenital system of rat fetuses. *Anat. Record* **139**: 145, 1961.
7. SHOTTON, D.; MONIE, I. W. Possible teratogenic effect of chlorambucil on a human fetus. *J.A.M.A.* **188**: 74, 1963.
8. SOKAL, J. E. AND LESSMAN, E. M. Effects of cancer chemotherapeutic agents on the human fetus. *J.A.M.A.* **172**: 1765, 1960.
9. MURPHY, M. L. AND CHAUBE, S. Teratogenic effects of abnormal purines and their ribosides. *Proc. Am. Assoc. Ca. Res.* **3**: 347, 1962.
10. THIERSCH, J. B. The effect of 6-mercaptopurine on the rat fetus and on reproduction of the rat. *Ann. N.Y. Acad. Sci.* **60**: 221, 1954.
11. MURPHY, M. L. Teratogenic effects of tumor-inhibiting chemicals in the foetal rat. *Ciba Foundation Symposium on Congenital Malformations*. London: Churchill, Ltd., 1960, pp. 78–114.
12. MURPHY, M. L. Teratogenic effects in rats of growth-inhibiting chemicals including thalidomide. *Clinical Proceedings of the Children's Hospital of Washington, D.C.*, **18**: 307, 1962.
13. FRIEDMAN, M. H. Personal communication, 1959.
14. NISHIMURA, H., DAGG, C. P., AND MURPHY, M. L. To be published.
15. CHAUBE, S. AND MURPHY, M. L. To be published.

DOSE-RESPONSE RELATIONSHIPS IN GROWTH-INHIBITING DRUGS IN THE RAT: TIME OF TREATMENT AS A TERATOLOGICAL DETERMINANT

EXPERIMENTAL PROCEDURE

Unmated Wistar rats weighing 180 to 200 g are caged with males at 10 P.M. Rats observed in copulation are isolated and considered zero days pregnant. At various intervals afterward, they are subjected to experimental procedures. They are shipped to the laboratory on about the fifth day. They are maintained in individual cages, fed a diet of Purina Chow with oatmeal, fresh vegetables, and water *ad libitum.*

Over 80 per cent of the control rats have had viable young at the time of littering, and 90 per cent of all of the implantation sites have had living fetuses. Experimental groups of pregnant animals are weighed and treated with intraperitoneal injections of appropriate doses of study drugs. Rats are held with a heavy glove to enlarge the hand and with the thumb and fore-finger on either side of the skull to prevent bites. Fingers around the neck, stretching the skin tightly, or anesthesia which might cause cyanosis are avoided. Injection is made with a 21-gauge needle into the right lower quadrant of the abdominal cavity. Rats are observed frequently on the day of treatment for any reaction to the chemicals and are weighed daily. Mesh cage bottoms are used to prevent rats from ingesting excreted drugs. Any rats which die are autopsied, and special attention is directed to observing appearance of the contents of the uterus. Controls weigh 4.9 ± 0.6 g.

On the twenty-first day of gestation, the day prior to expected littering for controls, the rats are anesthetized with chloroform (ether or pentobarbitol sodium are satisfactory) and the abdomen is opened. The uteri of control and treated rats are carefully examined. The total number of living and dead fetuses and implantation sites is regarded as the total number of conceptuses. Viable embryos are weighed and examined grossly. If they are to be studied later, the abdomens are opened and stored in 10 per cent neutral formalin or Bouin's solution. Usually, they are eviscerated, stored in 95 per cent ethanol, and later cleared and stained for study of the osseous skeleton.

Chemicals are dissolved in water. Insoluble ones are suspended in 0.5 per cent carboxymethylcellulose (CMC). The injected volume is constant at 0.01 cc/gram of rat. Chemicals are prepared just prior to injection and not stored in solution.

Procedure for Staining Rat Skeleton with Alizarin Red

Dr. S. Chaube, Sloan-Kettering Institute

Fixative.—95 per cent ethanol (4–7 days). Remove all of the abdominal and thoracic organs and as much of the skin as possible. This may be done at the time of sacrifice or after a few days in fixative.

Clearing.—Clear in 2 per cent KOH for 6–8 hours until bones are clearly visible through the surrounding tissue. Avoid excessive temperature elevation in the laboratory as soft tissues may disintegrate rapidly. Small specimens must be watched carefully and removed to avoid their disintegration.

Staining.—Transfer directly to Mall's solution[1] containing alizarin red (approximate proportion 100 ml of Mall's with 2–3 drops of saturated aqueous solution of alizarin red). Leave specimen overnight in stain.

Differential decolorization.—Transfer stained specimens to Mall's next morning. Make frequent changes to hasten the process, and observe until the bones are clearly seen through the muscles.

Upgrading in glycerine.—Transfer through several changes of 50, 70, 80, and 90 per cent glycerine in water. Examine and store in 100 per cent glycerine. Add a few crystals of thymol to prevent fungal growth.

Cartilage Staining

Drs. Asling and Monie, Department of Anatomy, University of California, San Francisco, California

Step 1.—10 per cent formalin—24 hours or longer.

Step 2.—eviscerate and skin—either submerge the embryo or frequently dip in formalin to prevent drying.

Step 3.—70 per cent ethyl alcohol + 1 per cent HCl. (99 cc 70 per cent alcohol + 1 cc conc. HCl). Any length of time.

Step 4.—1 per cent methylene blue in acid alcohol—2 days (19/100 cc).

Step 5.—destain in acid alcohol. Two or three changes, 24 hours each. Continue until very little or no dye is washed out.

Step 6.—95 per cent alcohol—24 hours or longer.

Step 7.—absolute ethyl alcohol—24 hours.

Step 8.—absolute alcohol + benzene (75:25)—24 hours.

Step 9.—absolute alcohol + benzene (50:50)—24 hours.

Step 10.—methyl salicylate.

Name.—

HYDROXYUREA

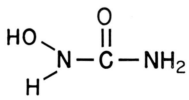

History.—Synthesized in 1869.

Chemistry.—Colorless weakly acid crystalline solid with a melting point of 140° C. It is readily soluble in water or ethyl acetate.

Biological activity.—At 1 M, hydroxyurea inhibits the action of urease. It is inactive in tissue culture.[2] Dr. Aaron Bendich has recently reported main-chain cleavages in DNA as measured by decrease in viscosity and loss of cytosine as with hydroxylamine. He also describes chromosome breakages, rearrangements, translocations, and uncoilings.

Toxicity in animals.[2]—Single dose toxicity IP non-pregnant animals. Mice: LD_{50} 7.5 g/kg; rats; LD_{50} 4.7 g/kg. Comment: Toxic doses within 15 to 30 min produce excitation followed by sedation, ataxia, tremors, respiratory arrest, convulsions, rigor, and death within 1–4 hours. Repeated dose toxicity IP or IV to non-pregnant animals: Rats: 50 mg/kg/day IP $\times$ 28. Four of 10 showed depression of marrow, one lymphoid depletion of spleen and infection. 200 mg/kg/day *per os* $\times$ 28. Three of 10 showed weight loss and pallor at death. Lungs were mottled and one had duodenal bleeding. Dogs: 200 mg/kg/day IV $\times$ 28. Marrow depression, depletion of lymphoid follicles of spleen, moderate spermatogenic arrest, and pneumonia occurred.

Tumor inhibitory effects.[2]—Mouse: Sarcomas, carcinomas, and leukemias were inhibited at IP doses of 100 to 167 mg/kg/day. Rat: Dunning leukemia was inhibited by *per os* doses of 400 mg/kg/day $\times$ 12. Comment: Hydroxyurea is a compound which when given orally, parenterally, or IV is active in a variety of tumor systems. It is also toxic by all routes and exerts its main effects on bone marrow, lymphoid tissue, lungs, and to some extent germinal elements and the gastrointestinal tract. That it also has neurotoxic effects is evidenced by the anesthetic and/or excitatory effects seen in all species studied. The provocative contrast between the massive single doses and the much smaller repeated doses necessary to produce toxicity is suggestive of rapid clearance of the drug from the body.

Teratogenic effects.—Sand dollar embryo: 200 gamma/ml blocks development at the morula stage. (Dr. D. Karnofsky and Miss E. Simmel.) Chick embryo: The LD_{50} dose was 0.4 to 0.5 mg/egg. Some which died before the eighteenth day showed limb defects. (Dr. Karnofsky and Mrs. C. Lacon.) Rat fetuses: On the ninth to twelfth days of gestation they showed resorption and abnormalities at doses of 500 to 2000 mg/kg IP to the pregnant rat.

Clinical trials.—Oral doses of 20 mg/kg/day have been used to control chronic myelogenous leukemia for several months. Doses of 40 to 80 or 100 mg/kg/day *per os* have produced leukopenia in patients with acute leukemia and metastatic solid tumors without objective improvement in the neoplastic condition.

[2] Data supplied by the Cancer Chemotherapy National Center of the National Cancer Institute.

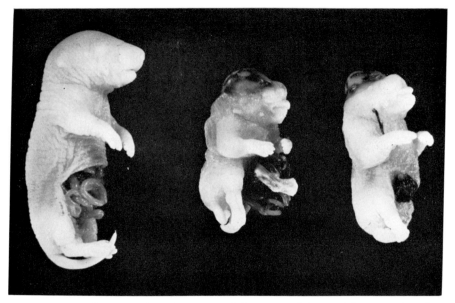

PLATE 1.—Control fetus on left and two treated fetuses on right at 21-days gestation. Pregnant animal was treated with 250 mg/kg of hydroxyurea, intraperitoneally on day 9 of gestation. The litter contained 13 fetuses and 11 were alive and 2 resorbed. Eight were abnormal. Gross malformations included exencephaly, cleft palate, retarded mandibles, retarded and clubbed rear leg, and coiled tail.

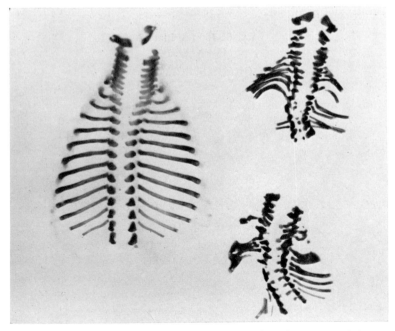

PLATE 2.—Fetal skeletal preparations showing control at left and two treated fetuses on right at 21 days of gestation. Pregnant animal was treated with 250 mg/kg of hydroxyurea intraperitoneally on day 9 of gestation. Litter contained 13 fetuses and 11 were alive and 2 resorbed. Eight were abnormal. Anomalies were severely damaged ribs and associated thoracic vertebrae and also fused fragmented cervical vertebrae.

GROSS SPECIMENS AND SKELETONS

Hydroxyurea (Diluent: Physiological Saline)

250 mg/kg intraperitoneal (single dose)

Treated . . . . . Day 9	Total implantation sites = 13	
Sacrificed . . . . . Day 21	No. resorbed 2	
Average Weight . . 3.9 g	No. living 11	
	No. abnormal . . . 8	

GROSS MALFORMATIONS

Exencephaly, cleft palate, retarded mandibles, retarded and clubbed rear leg, coiled tail.

RAT 63-422 f 1, Plate 2

SKELETAL ABNORMALITIES

Note absence of the frontals, parietals, interparietal, and supraoccipital bones in the skull, fused irregular cervical vertebrae, and fused ribs.

RAT 62-460 f 2

SKELETAL ABNORMALITIES

Ribs.—Another specimen from the same litter which shows more severely damaged ribs and associated thoracic vertebrae; also shows fused and fragmented cervical vertebrae.

Skull.—Note the following dorsal skull bones are completely absent: frontals; parietals; supraoccipitals; and interparietal. The following are partially present: maxillae; palatines; and pterygoids.

RAT 62-460 f 3, Plate 2

GROSS SPECIMEN

Hydroxyurea (Diluent: Physiological Saline)

500 mg/kg intraperitoneal (single dose)

Treated . . . . . Day 11	Total implantation sites = 10
Sacrificed . . . . . Day 21	No. resorbed 0
Average weight . . . 3.7 g	No. living 10
	No. abnormal . . . 10

GROSS MALFORMATIONS

Cleft palate, clubbed rear leg, ectrodactylous rear paw, retarded and curled tail.

RAT 62-357 f 4

GROSS SPECIMEN

Hydroxyurea (Diluent: Physiological Saline)

750 mg/kg intraperitoneal (single dose)

Treated . . . . . Day 12	Total implantation sites = 14
Sacrificed . . . . . Day 21	No. resorbed 1
Average weight . . . 2.9 g	No. living 13
	No. abnormal . . . 13

M. Lois Murphy 166

Retarded mandibles, clubbed fore leg, and polydactylous rear paws.

Gross specimens had bloody patches on the head. In cleared and stained (alizarin red) specimens, the bones in this area (occipitals and parietals) were incompletely ossified.

RAT 64-15 f 5

GROSS SPECIMEN

Hydroxyurea (Diluent: Physiological Saline)

1000 mg/kg intraperitoneal (single dose)

Treated . . . . . Day 12	Total implantation sites = 10	
Sacrificed . . . . . Day 21	No. resorbed 3	
Average weight . . . 2.2 g	No. living 7	
	No. abnormal . . . 7	

GROSS MALFORMATIONS

Exencephaly, retarded mandible, retarded and clubbed fore and rear legs, ectrodactylous fore paw, splayed rear paw, and retarded tail.

RAT 62-327 f 6

GROSS SPECIMENS—ENTIRE LITTER

Hydroxyurea (Diluent: Physiological Saline)

925.0 mg/kg intraperitoneal (total dose)—
175, 250, and 500 mg/kg

Treated . . . . . Days 10, 11, and 12	Total implantation sites = 11
Sacrificed . . . . . Day 21	No. resorbed 2
Average weight . . . 2.0 g	No. living 9
	No. abnormal . . . 9

GROSS MALFORMATIONS

No. Fetuses	Specific Abnormalities
1	Snout repressed, mild encephaly, cleft palate, retarded mandibles, retarded and clubbed fore and rear legs, tail absent.
1	Snout repressed, mild encephaly, cleft palate, cleft lip (right), retarded mandibles, retarded and clubbed fore and rear legs, retarded and curled tail.
1	Snout repressed, encephaly, retarded mandibles, retarded and clubbed fore and rear legs.
2	Mild encephaly, retarded mandibles, retarded and clubbed fore legs, retarded and curled tail.
4	Snout repressed, mild encephaly, cleft lip (bilateral), retarded mandibles, retarded and clubbed rear legs.

RAT 63-530 f 7 HU

SELECTED BIBLIOGRAPHY

CHAUBE, S., SIMMEL, E., AND LACON, C. Hydroxyurea, a teratogenic chemical. *Proceedings of the American Association for Cancer Research.* 4(1):10, 1963.

DAVIDSON, J. D. AND WINTER, T. S. A method of analyzing for hydroxyurea in biological fluids. *Cancer Chemo. Reports.* **27**: 97, 1963.

KRAKOFF, I. H., MURPHY, M. L., AND SAVEL, H. Preliminary trials of hydroxyurea in neoplastic diseases in man. *Proceedings of the American Association for Cancer Research.* 4(1): 35, 1963.

LERNER, L. J., BIANCHI, A., DZELZKALNS, M., AND DePHILLIPO, M. Antileukemia and antitumor activities of hydroxyurea in the experimental animal. *Proceedings of the American Association for Cancer Research.* 4(1): 37, 1963.

MURPHY, M. LOIS AND CHAUBE, S. Preliminary survey of hydroxyurea as a chemical teratogen. *Cancer Chemo. Reports,* **40:** 1, 1964.

THURMAN, WILLIAM G. Pharmacology and antitumor effect of hydroxyurea. *Proceedings of the American Association for Cancer Research.* 4(1): 67, 1963.

Name.—

HADACIDIN

$$H-\overset{\overset{O}{\|}}{C}-\overset{\overset{OH}{|}}{N}-CH_2-\overset{\overset{O}{\|}}{C}-OH$$

N-Formyl-N-hydroxyglycine

History.—Synthesized in 1962. Identified from broth culture of *Penicillium frequentans.*

Chemistry.—N-Formyl-N-hydroxyglycine: molecular weight = 141; melting point = 119° C; water soluble; insoluble in ethyl alcohol.

Biological activity.—It is a potent inhibitor of adenylosuccinic synthetase prepared from *E. coli* and it is competitively reversed by L-aspartate. There was no inhibition of anaerobic bacteria. Amounts of other growth inhibiting drugs required to inhibit growth of *E. coli* are reduced by 1000 gamma/ml of Hadacidin and >100 gamma/ml inhibited by 50 per cent *Streptococcus faecalis.*

Toxicity in animals.—Single dose toxicity IP in non-pregnant animals. Mice: 10 g/kg within 10–15 min, hyperactivity, gross tremors, apnea, and terminal tonic convulsions. All died within 30 mins. 5 g/kg within 10–15 min. Those alive after 2 hours were alive 24 hours later showing only weight loss. Rats: 5 and 10 g/kg, 5/6 at each dose died. Repeated dose toxicity IV or IP to non-pregnant animals. Mice: 2.5 g/kg/day × 5. No effects. Rats: 2.5 g/kg/day × 5. No effects. Dogs: 1 g/kg/day × 10. Each injection was followed by slight depression and ataxia which disappeared within 15 min. Some animals had vomiting or diarrhea, polydypsia and polyuria which ceased after the drug was stopped. No blood, biochemical, or hematological abnormalities were seen. (Dr. F. Phillips).

Tumor inhibitory effects.—Human intestinal carcinoma HAD1; human bronchogenic carcinoma A42, both growing in embryonated eggs were inhibited by Hadacidin at 5 or 6 mg/egg. It did not, however, inhibit human

M. Lois Murphy   168

sarcoma 1, growing in the hamster cheek pouch or several transplantable rodent tumors growing in mice or rats at doses of 2 g/kg/day × 7.

Teratogenic effects.—Chick: 2 to 4 mg/egg was LD$_{50}$. At 2 mg/egg on day 4, abnormalities of beak, limbs, and eyes were seen (Dr. Karnofsky). Rats: Doses of 2.5 g/kg at 6-hr intervals or repeated on 2 or more days gave abnormalities in fetuses with no toxicity in the adult.

Clinical trials.—Cancer patients receiving oral doses of 2.5 to 60 g/day showed vomiting, diarrhea, and possibly leucopenia. In more recent studies, however, when the drug was given at 30 mg/kg every 6 hours, all patients developed signs of marrow depression and stomatitis. No specific benefit to cancer patients has yet been demonstrated but clinical trials with lower fractionated doses continue.

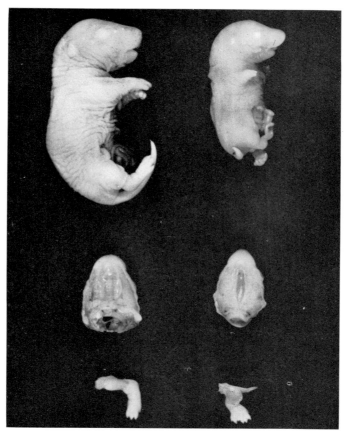

PLATE 3.—Fetuses from pregnant rats treated with 6-hr fractionated doses of Hadacidin. Fetuses were sacrificed on the twenty-first day of gestation. Left is control and right is fetus treated with 5.0 gm/kg total at 2.5 gm/kg × 2, 6-hr fractionated dose on the twelfth day of gestation. Palates and left rear appendages of littermates appear vertically below each fetus. The palates were exposed by removing the lower jaw and the tongue. There are cleft palate, retarded mandibles, retarded and clubbed fore and rear legs, ectrodactylous fore paw, polydactylous rear paw, and retarded kinky tail. Total implantation sites were 14 with 7 living, 7 resorbed, and 7 abnormal.

169   *Time and Dose-Response Relations*

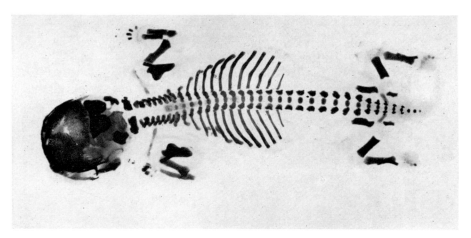

PLATE 4.—Dorsal view of a control fetus. Fetal skeleton at 21 days.

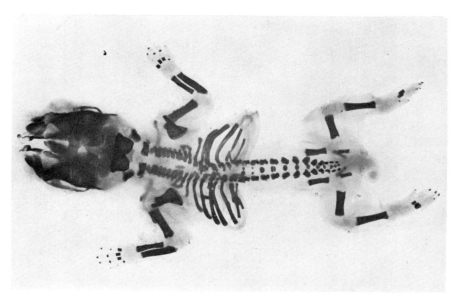

PLATE 5.—Dorsal view of a fetus from a pregnant rat treated with 2.5 gm/kg × 2, 6-hr fractionated dose Hadacidin on the tenth day of gestation. Total implantation sites were 12 with 7 living, 5 resorbed, and 3 abnormal. Note the absence of the major portions of the dorsal and ventral cranial bones comprising the anterior two-thirds of the head and the fused, thickened, and crooked ribs, angulated scapula, fused, irregular (thoracic and lumbar). and absent (caudal) vertebrae.

GROSS SPECIMEN

Hadacidin   (Diluent: Carboxymethylcellulose)

5000 mg/kg intraperitoneal (Fraction-   Total Implantation sites = 12
    ated—6 hours)
Treated . . . . . Day 10   No. resorbed 3
Sacrificed . . . . . Day 21   No. living 6
Average weight . . . 3.0 g   No. abnormal . . . 6
                                           No. dead 3

GROSS MALFORMATIONS

Exencephaly, cleft palate, cleft lip, retarded mandibles, clubbed fore leg, ectrodactylous fore paw.

RAT 90 f 7

FETAL SKELETON

Hadacidin   (Diluent: Carboxymethylcellulose)

5000 mg/kg intraperitoneal (Fraction-   Total implantation sites = 12
    ated—6 hours)
Treated . . . . . Day 10   No. resorbed 5
Sacrificed . . . . . Day 21   No. living 7
Average weight . . . 3.2 g   No. abnormal . . . 3

SKELETAL ABNORMALITIES

Absence of anterior ⅔ of the frontal bones and incomplete ossification of the nasal bones in the skull; absent and/or fused ribs; retarded and fused thoracic vertebrae.

RAT 50 f 8, Plates 4 and 5

GROSS SPECIMEN AND SKELETON

Hadacidin   (Diluent: Carboxymethylcellulose)

5000 mg/kg intraperitoneal (Fraction-   Total implantation sites = 8
    ated—6 hours)
Treated . . . . . Day 11   No. resorbed 6
Sacrificed . . . . . Day 21   No. living 2
Average weight . . . 2.7 g   No. abnormal . . . 2

GROSS MALFORMATIONS

Cleft palate, cleft lip, retarded mandibles, clubbed rear leg, ectrodactylous fore and rear paw, retarded and kinky tail.

RAT 53 f 9

SKELETAL ABNORMALITIES

Incomplete ossification of the bones of the skull (frontals and parietals), retarded curved radius and ulna; angulated scapula; angulated and retarded ribs; fusion of vertebral arches (especially in the lumbar region); and fused and distorted vertebral centra (in both thoracic and lumbar regions).

RAT 53 f 10

171   *Time and Dose-Response Relations*

GROSS SPECIMEN

Hadacidin   (Diluent: Carboxymethylcellulose)

5000 mg/kg intraperitoneal (Fraction-   Total implantation sites = 14
ated—6 hours)

Treated . . . . .   Day 12   No. resorbed   7
Sacrificed . . . . .   Day 21   No. living   7
Average weight .   . .   2.5 g   No. abnormal   . . .   7

GROSS MALFORMATIONS

Cleft palate, retarded mandibles, retarded and clubbed fore and rear legs, ectrodactylous fore paw, polydactylous rear paw, and retarded, kinky tail.

RAT 4831 f 11, Plate 3

GROSS SPECIMEN AND SKELETON

Hadacidin   (Diluent: Carboxymethylcellulose)

2500 mg/kg × 3 intraperitoneal (Multiple)   Total implantation sites = 13

Treated . . . .   Days 9, 10, and 11   No. resorbed   5
Sacrificed . . . .   Day 21   No. living   8
Average weight .   .   2.6 g   No. abnormal   . . .   8

GROSS MALFORMATIONS

Exencephaly, retarded, absent nasal and frontal regions of head, depressed occipital area, absent maxillae, retarded mandibles, eyes absent, retarded fore leg.

RAT 4964 f 12

SKELETAL ABNORMALITIES

Incomplete ossification of dorsal bones (frontals, parietals, and interparietal) of the skull; curved radius and ulna; angulated scapula and spine; fusion of the last cervical vertebra with the first thoracic; fused thoracic vertebrae and fused ribs. The anterior ⅔ of the skull of a littermate is so severely damaged as to show complete absence of the following bones—premaxillae, maxillae, nasals, ethmoids, palatines, frontals, and parietal.

RAT 4964 f 13

GROSS SPECIMEN

Hadacidin   (Diluent: Carboxymethylcellulose)

2500 mg/kg × 3 intraperitoneal (Multiple)   Total implantation sites = 11

Treated . . . . .   Days 10, 11, and 12   No. resorbed   9
Sacrificed . . . . .   Day 21   No. living   1
Average weight .   . .   2.2 g   No. abnormal   . . .   1
                                             No. dead   1

GROSS MALFORMATIONS

Maxillae absent, cleft palate, cleft cheek, retarded mandibles, retarded fore leg, retarded and clubbed rear leg, ectrodactylous (right) and polydactylous (left) paws, and retarded tail.

RAT 4969 f 14

M. Lois Murphy   172

SELECTED BIBLIOGRAPHY

CHAUBE, S. AND MURPHY, M. L. Teratogenic effect of Hadacidin (a new growth inhibitory chemical) on the rat fetus. *Journal of Experimental Zoology.* **152**: 67, 1963.

ELLISON, R. R. Preliminary clinical trials of Hadacidin, a new tumor-inhibitory substance. *Clinical Pharmacology and Therapeutics.* **4**: 326, 1963.

GITTERMAN, C. O., DULANEY, E. L., KACAKA, E. A., HENDLIN, D., AND WOODRUFF, H. B. The human tumor-egg host system. II. Discovery and properties of a new antitumor agent, Hadacidin. *Proc. Soc. Exper. Biol. Med.* **109**: 852, 1962.

WHITE, F. R. Hadacidin, data compiled for CCNSC. *Ca. Chemo. Reports,* **23**: 81, 1962.

Name.—5-FLUOROURACIL

Chemistry.—Analogue of uracil.

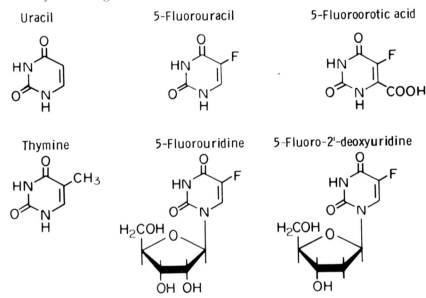

History.—The synthesis and biological activity of 5-fluorouracil was first reported in 1957. Subsequently, a large number of fluorinated pyrimidine analogues were synthesized and tried as potential anticancer drugs: 5-fluoroorotic acid (FO); 5-fluorouridine (FUR), and 5-fluoro-2'-deoxyuridine (FUDR).

Biological activity.—The fluorinated pyrimidines alter cell function by two main actions: One effect is on RNA and DNA synthesis; FU and FUR can be incorporated into RNA as abnormal nucleotides and also inhibit incorporation of orotic acid and uracil into RNA. FU may also block enzyme synthesis or alter RNA and protein synthesis to form abnormal enzymes. The second and perhaps most important action is inhibition of thymidylate synthetase (FUDRMP). This enzyme catalyzes the methylation of 2'-deoxyuridylic acid (UDRMP) to form thymidylic acid (de novo thymidine synthesis). The action of 5-bromo-, 5-iodo-, or 5-chloro-deoxyuridine in contrast interferes with incorporation of thymidine itself into DNA and in fact may be incorporated instead of thymidine into DNA of cells.

Signs of chronic intoxication were weight loss and diarrhea with bleeding. Pathological studies revealed hemorrhages in the colon, depletion of lymphoid tissues, atrophic epithelial tissues with hemorrhage, and depletion of nucleated cells in bone marrow.

Tumor inhibitory effects.—These agents are active against rodent tumors.

Teratogenic effects.—These agents have been studied in the sand dollar embryo, chick, mouse, and rat fetuses and have shown activity in all.

ACUTE TOXICITY INTRAPERITONEAL LD$_{50}$ MG/KG (1 DOSE)[a, b]

Agent	Mice	Rats	Cats
FU	365	230	12–25
FO	430	>500	25–50
FUR			12–25
FUDR		>1,000<2,000	12–25
FCDR		>2,000	12–25

[a] The acute effects seen in cats were convulsive seizures within a few hours and death within 24 hours. Others recovered in 2 to 4 days.

[b] From F. Philips.

SUCCESSIVE DOSE TOXICITY LD$_{50}$ MG/KG/DAY$\times$5[a]

Agent	Mice	Rats	Dogs[a]	Rhesus Monkeys[b]
FU	50	25	7	20
FO	32	50	3	
FUR	10	43	1	1.3
FUDR	210	70	20	40
FCDR	300	5.0	4	40

[a] From F. Philips.

[b] Dose $\times$ 10.

Clinical trials.—They all have been tried in patients with cancer. Those with the most extensive trials are FU and FUDR. The former tried at doses of 15 mg/kg/day intravenously $\times$ 5 or the latter at 30 mg/kg/day intravenously $\times$ 5 produces mild to severe toxicity, stomatitis, diarrhea, and hypoplastic marrow along with variable regression of metastatic tumors. Iododeoxyuridine has been instilled locally into the eye in management of keratoconjunctivitis from herpes simplex.

GROSS SPECIMENS—ENTIRE LITTER

5-Fluorodeoxyuridine   (Diluent: Distilled Water)

50 mg/kg intraperitoneal (single dose)   Total implantation sites = 10

Treated . . . . . Day 9   No. resorbed 2

Sacrificed . . . . . Day 21   No. living 8

Average weight . . . 4.0 g   No. abnormal . . . 4

M. Lois Murphy   174

TABLE 1

CHEMOTHERAPEUTIC AGENTS USED IN TREATING HUMAN CANCER

Comparison of Clinical Dosages with Those Producing Lethal or Teratogenic Effects in Rat Fetuses

Agent	Adult Rat — Est. LD$_{50}$ (mg/kg/day) 1 dose	Est. LD$_{50}$ 5 doses	Pregnant Rats 11th or 12th Day of Gestation — Est. Fetal Min. LD$_{100}$ (mg/kg/day) 1 dose	Teratogenic Effects in Fetuses (mg/kg/day) 1 dose	Fetal Maximum Non-teratogenic (mg/kg/day) 1 dose	Therapeutic Use in Man — Dosage (mg/kg/day)	No. Days
Alkylating agents:							
Nitrogen mustard	2.0	......	1.5	0.5 to 0.7	0.25	0.4	1
TEM	1.25	......	1.0	0.5 to 0.75	0.3	0.04	×3
Thio-TEPA	8.0	......	6.0	3.0 to 5.0	1.5	0.2	×5
Chlorambucil	24.0	......	18.0	6.0 to 10.0	3.0	0.1 to 0.2	daily
Myleran	60.0	......	45.0	18.0 to 34.0	12.0	0.06	daily
Cytoxan	40.0	......	20.0	15.0 to 20.0	10.0	2.0 to 5.0	daily
Antimetabolites.—							
Folic acid antagonists:						mg/day	
Methotrexate	17.0	2.5	2.5	none observed	0.6	2.5	daily
Aminopterin	3.4	0.5	0.5	none observed	0.1	0.5	daily
Purine analogues:						mg/kg/day	
6-MP	250	......	125.0	31.0 to 100.0	15.0	2.5	daily
TG	350	......	25.0	12.0 to 50.0	6.0	2.5	daily
Pyrimidine analogues:							
FU	230	25	50.0	12.0 to 37.0	10.0	15.0	×5
Miscellaneous:							
Actinomycin D	0.4	0.088	0.3	none observed	0.2	0.015	×5
Vincristine	2.0	......	1.25	none observed	0.65	0.05–0.1	×1/wk

Nitrogen mustard = methyl-*bis* (β-chlorethyl)amine hydrochloride; TEM = triethylene melamine; Thio-TEPA = triethylenethiophosphoramide; Myleran = 1,4-dimethanesulfonyloxybutane; Cytoxan = cyclophosphamide; Methotrexate = 4-amino-N^{10}-methyl-pteroylglutamic acid; Aminopterin = 4-amino-pteroylglutamic acid; 6-MP = 6-mercaptopurine; TG = thioguanine; FU = 5-fluorouracil.

TABLE 2

CHEMOTHERAPEUTIC AGENTS WITHOUT ESTABLISHED USE IN HUMAN CANCER

Comparison of Clinical Dosages with Those Producing Lethal or Teratogenic Effects in Rat Fetuses

ANTIMETABOLITES	ADULT RAT Est. LD$_{50}$ (mg/kg/day)		PREGNANT RATS 11TH OR 12TH DAY OF GESTATION			THERAPEUTIC TRIAL IN MAN	
	1 dose	5 doses	Est. Fetal Min. LD$_{100}$ (mg/kg/day) 1 dose	Teratogenic Effects in Fetuses (mg/kg/day) 1 dose	Fetal Maximum Non-teratogenic (mg/kg/day) 1 dose	Dosage (mg/kg/day)	No. Days
Glutamine antagonists:							
DON	80	0.3	0.2	none observed	0.05	1.0	7–10
Azaserine	100	25	5.0	2.0 to 3.0	1.0	10	7–10
Purine analogues:							
Purine	800–1,000	......	800	none observed	400	......	
6MPR	2,000–3,000	......	250	31–150	15	2.0	daily
TGR	200–400	......	25	12–25	6.2	2.0	daily
9-Ethyl-6-mercaptopurine	500–600	......	>500	400–500	250	5–10	daily
9-Butyl-6-mercaptopurine	300–500	......	300	300	<100	......	
Pyrimidine antagonists:							
FO	300–400	50	300	150	100	1–3	10–15
FUR	400–800	43	>100	25	12	0.75	3
FUDR	1,600	70	200	50–150	25	30	5
FCDR	>2,000	5	5.0	0.15–2.5	<0.15	......	
BrCDR	>1,500	......	......	1,500	......	......	
IUDR	4,000–8,000	......	>2,500	1,250–2,500	1,000	100	5
ClUDR	>2,000	......	......	1,000	......	......	
BrUDR	1,500–4,000	......	1,000	500–800	250	......	
Nicotinamide antagonists:							
6-aminonicotinamide	20	......	15	5–8	2.5	5	5–40
5-fluoronicotinamide	560	150	200	none observed	70	......	

TABLE 2—Continued

MISCELLANEOUS	ADULT RAT Est. LD₅₀ (mg/kg/day)		PREGNANT RATS 11TH OR 12TH DAY OF GESTATION			THERAPEUTIC TRIAL IN MAN	
	1 dose	5 doses	Est. Min. LD₁₀₀ (mg/kg/day) 1 dose	Teratogenic Effects in Fetuses (mg/kg/day) 1 dose	Fetal Maximum Non-teratogenic (mg/kg/day) 1 dose	Dosage (mg/kg/day)	No. Days
Thiadiazole	200	.....	200	50–200	25	1–3	14–20
Triazene	180	.....	.....	30	15	8	14+
Actinobolin	.....	.....	2,000	none observed	500	50–100	1–30
Mitomycin	2.0–2.5	.....	2.0–2.5	none observed	1.5	0.2	×3
Hadacidin	<5,000	.....	>2,000	2,500–5,000	1,250	70	q6h
P1875	.....	.....	150	50	35	20–50	.....
8-Azaguanine	.....	.....	>500	500	300	.....	.....
Hydroxylamine	100–125	.....	>62 <125	none observed	62.5	none	.....
Hydroxyurea	>2,000	.....	1,000–2,000	500–750	250	20	20–30
Hydroxyurethane	750	.....	750–1,000	500–700	250	.....	.....
Urethane	1,500	.....	1,000–1,500	none	800	.....	.....
Isometamidium	25	.....	50	none	25	1–2	10–20
Methyl-glyoxal-bis-guanylhydrazone	100–150	.....	.....	none observed	.....	5–10	.....
1,3-bis(2-Chloroethyl)1-nitrosurea	>20	.....	>150	none	10	3	21+
Methylhydrazine	800	.....	200–300	50–300	12	5–10	×1/week

DON = diazo-oxo-L-norleucine
Azaserine = O-diazoacetyl-L-serine
6-MPR = 6-mercaptopurine riboside
TGR = thioguanine riboside
FO = 5-fluoroorotic acid

FUR = 5-fluorouridine
FUDR = 5-fluorodeoxyuridine
FCDR = 5-fluorodeoxycytidine
BrCDR = 5-bromo-deoxycytidine
IUDR = 5-iodo-2'-deoxyuridine

ClUDR = 5-chlorodeoxyuridine
BrUDR = 5-bromodeoxyuridine
Thiadiazole = 2-ethylamino-1,3,4-thiadiazole
Triazene = 3,3-dimethyl-1-phenyltriazene
Hadacidin = n-formyl hydroxyaminoacetic acid

TABLE 3

MATERNAL AND FETAL TOXICITY AND TERATOGENIC EFFECTS WITH FIVE FLUORINATED PYRIMIDINES

Drug	FU	FO	FUR	FUDR
Maternal LD$_{50}$: Dose (mg/kg)	230	300–400	400–800	800–1,600
Dosage range causing fetal effects (mg/kg)	25–100	50–300	25–100	12–100

MATERNAL AND FETAL TOXICITY AND TERATOGENIC EFFECTS WITH FIVE FLUORINATED PYRIMIDINES

DOSAGES AFFECTING FETUSES AT VARIOUS DAYS OF GESTATION (mg/kg to pregnant rats IP)

DAYS OF GESTATION	9	10	11	12	14	9	10	11	12	14	9	10	11	12	14	9	10	11	12	14	17
100% Fetal resorption	50	50	37.5	50	>100	50	100	150	>300	.....	75	100	>100	>100	>100	75	100	100	100	>100	>100
Teratogenic	25	25	25	25	100	25	.....	100	150	200	25	12	25	25	100	25	12	25	25	100	100
Sub-teratogenic dose	25	5	15	10	.....	25	50	.....	100	.....	<25	75	<25	12	<100	<25	<12	<25	12	<100	<100

MATERNAL AND FETAL TOXICITY AND TERATOGENIC EFFECTS WITH FIVE FLUORINATED PYRIMIDINES

(Maximum Tolerated Dosages in Rats with Transplanted Tumors (mg/kg (after Sugiura)))

	FU	FO	FUR	FUDR
Daily dose	25	20	2.5	40
Total course	175	140	17.5	280
Inhibition	.....	.....	.....	.....

Exencephaly, cleft palate, cleft lip, cleft face, eyes absent, ears retarded.

RAT 3269 f 15, Plate 6

GROSS SPECIMENS—ENTIRE LITTER

5-Fluorodeoxyuridine (Diluent: Distilled Water)

75 mg/kg intraperitoneal (single dose)	Total implantation sites = 16	
Treated . . . . . Day 10	No. resorbed 1	
Sacrificed . . . . . Day 21	No. living 15	
Average weight . . . 3.6 g	No. abnormal . . . 4	

GROSS MALFORMATIONS

Retarded and clubbed rear legs and ectrodactylous rear paws.
In skeletons stained with alizarin red, fused ribs and abnormal vertebrae were observed.

RAT 3252 f 16

PLATE 6.—This fetus is an example from a litter whose mother received 25 mg/kg of FUDR on the ninth day of gestation and was sacrificed on the twenty-first day. Fetuses in a litter are fewer than controls and are larger, weighing as much as 6 g as compared to the controls 4.9 ± 0.6 g. The lip, palate, and frontal skull are cleft; the tongue protrudes; there is anterior exencephaly. The liver and intestines protrude; the right leg is clubbed and the tail kinked.

GROSS SPECIMENS—ENTIRE LITTER

5-Fluorodeoxyuridine   (Diluent: Distilled Water)

75 mg/kg intraperitoneal (single dose)	Total implantation sites = 13	
Treated . . . . . Day 11	No. resorbed & dead . 3	
Sacrificed . . . . . Day 21	No. living 10	
Average weight . . . 3.8 g	No. abnormal . . . 2	

GROSS ABNORMALITIES

Clubbed fore leg and retarded tail.

RAT 3094 f 17

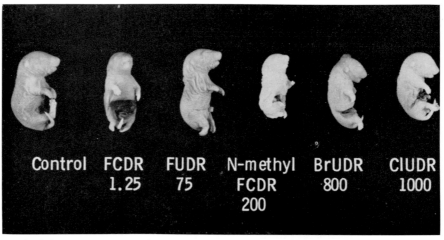

PLATE 7.—These are a control fetus on the left and five examples from litters whose mothers were treated on the twelfth day of gestation with the several halogenated pyrimidines and dosages in mg/kg intraperitoneally listed under each example. Deformed appendages, tails, and stunting are generalized. The one receiving bromodeoxyuridine has an encephalocele as well.

GROSS SPECIMENS—ENTIRE LITTER

5-Fluorodeoxyuridine   (Diluent: Distilled Water)

75 mg/kg intraperitoneal (single dose)	Total implantation sites = 7	
Treated . . . . . Day 12	No. resorbed 0	
Sacrificed . . . . . Day 21	No. living 7	
Average weight . . . 4.4 g	No. abnormal . . . 3	

GROSS MALFORMATIONS

Pointed nose, cleft palate, clubbed rear leg, polydactylous rear paw, and retarded tail.

RAT 3286 f 18

GROSS SPECIMENS—ENTIRE LITTER

5-Fluorodeoxyuridine   (Diluent: Distilled Water)

200 mg/kg intraperitoneal (single dose)	Total implantation sites = 11	
Treated . . . . . Day 14	No. resorbed 2	
Sacrificed . . . . . Day 21	No. living 9	
Average weight . . . 2.9 g	No. abnormal . . . 4	

M. Lois Murphy   180

Severe retardation of the body, pointed snout, retarded mandibles, retarded and clubbed fore and rear legs, ectrodactylous fore and rear paws, retarded tail and severe edema.

RAT 3148 f 19

GROSS SPECIMENS

5-Fluorodeoxycytidine (Diluent: Distilled Water)

1.25 mg/kg intraperitoneal (single dose)

Treated . . . . . Day 12	Total implantation sites = 11
Sacrificed . . . . . Day 21	No. resorbed 6
Average weight . . . 2.8 g	No. living 5
	No. abnormal . . . 5

GROSS MALFORMATIONS

Encephaly, cleft palate, retarded and clubbed fore and rear legs, ectrodactylous fore and rear paws, retarded, kinky tail.

RAT 198 f 20, Plate 7

GROSS SPECIMEN

5-Fluorodeoxyuridine (Diluent: Distilled Water)

75 mg/kg intraperitoneal (single dose)

Treated . . . . . Day 12	Total implantation sites = 11
Sacrificed . . . . . Day 21	No. resorbed 0
Average weight . . . 3.9 g	No. living 11
	No. abnormal . . . 10

GROSS MALFORMATIONS

Pointed nose, cleft palate, clubbed rear leg, polydactylous rear paw, and retarded tail.

RAT 32-83 f 21

GROSS SPECIMEN

N-Methyl-5-fluorodeoxycytidine (Diluent: Distilled Water)

200 mg/kg intraperitoneal (single dose)

Treated . . . . . Day 12	Total implantation sites = 13
Sacrificed . . . . . Day 21	No. resorbed 0
Average weight . . . 2.4 g	No. living 13
	No. abnormal . . . 13

GROSS MALFORMATIONS

Cleft palate, retarded and clubbed fore and rear legs, ectrodactylous rear paw, adactylous fore paw, and retarded tail.

RAT 62-87 f 22

GROSS SPECIMEN

5-Bromodeoxyuridine (Diluent: Distilled Water)

800 mg/kg intraperitoneal

Treated . . . . . Day 12	Total implantation sites = 15
Sacrificed . . . . . Day 21	No. resorbed 7
Average weight . . . 2.2 g	No. living 8
	No. abnormal . . . 8

Exencephaly, cleft palate, cleft lip, retarded and clubbed fore and rear legs, ectrodactylous and syndactylous fore or rear paws, and retarded tail.

RAT 6275 f 23, Plate 7

GROSS SPECIMEN

5-Chlorodeoxyuridine   (Diluent: Distilled Water)

1,000 mg/kg intraperitoneal (single dose)	Total implantation sites = 15
Treated . . . . . Day 12	No. resorbed 2
Sacrificed . . . . . Day 21	No. living 13
Average weight . . . 2.9 g	No. abnormal . . . 13

Mild exencephaly, cleft palate, retarded and clubbed fore and rear legs, ectrodactylous fore paw, polydactylous rear paw, and retarded tail.

RAT 62-51 f 24, Plate 7

GROSS SPECIMEN

5-Iododeoxyuridine   (Diluent: Distilled Water)

1,250 mg/kg intraperitoneal (single dose)	Total implantation sites = 12
Treated . . . . . Day 12	No. resorbed 11
Sacrificed . . . . . Day 21	No. living 1
Average weight . . . 2.0 g	No. abnormal . . . 1

Cleft palate, cleft lip, retarded, clubbed fore and rear legs, ectrodactylous fore paw, splayed and abnormal rear paws, and retarded tail.

RAT 4880 f 25

SELECTED BIBLIOGRAPHY

DAGG, C. P. Sensitive stages for the production of developmental abnormalities in mice with 5-fluorouracil. *Am. J. Anat.* **106:** 89, 1960.

DAGG, C. P. Sensitive stages for the production of developmental abnormalities in mice with 5-fluorouracil. *Cancer Chemo. Abstracts* **1:** 980, 1960.

DUCHINSKY, R., PLEVEN, E., AND HEIDELBERGER, C. Synthesis of 5-fluoropyrimidines. *J. Am. Chem. Soc.* **79:** 4559, 1957.

ELLISON, R. R. Clinical applications of the fluorinated pyrimidines. *Medical Clinics N.A.* **45:** 677, 1961.

KARNOFSKY, D. Cellular effects of anticancer drugs. *Ann. Rev. Pharma.* **3:** 357, 1963.

KAUFMANN, H. E. Clinical cure of herpes simplex keratitis by 5-iodo-2′deoxyuridine *Proc. Soc. Exper. Biol. Med.* **109:** 251, 1962.

PHILIPS, F. S. AND STERNBERG, S. S. Notes on the toxicity of 2′-deoxy-5-fluorocytidine and of other 5-fluoropyrimidines. *Informal Report.* Personal communication.

Name.—6-MERCAPTOPURINE

History.—Synthesized in 1952.

Biological activity.—Structural analogue of adenine and hypoxanthine. It is soluble as the sodium salt. It interferes with incorporation of preformed purines and pyrimidines into nucleic acids. 6-Mercaptopurine is readily converted to the ribotide, and it interferes with various mechanisms concerned with purine metabolism. The major effect appears to be related to a block in the interconversion of purines from inosinate at the nucleotide level. The development of resistance by cells is accompanied by a decrease in IMP pyrophosphorylase activity which decreases their ability to convert mercaptopurine to its nucleotide.

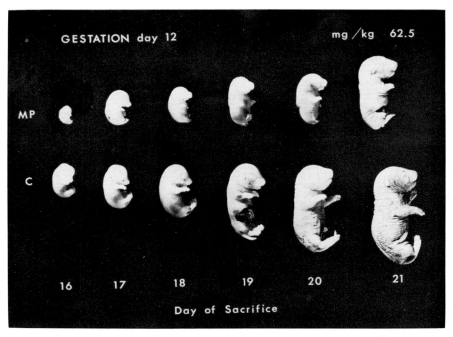

PLATE 8.—Fetuses from control and 6-mercaptopurine-treated rats on the twelfth day of gestation were sacrificed beginning on the sixteenth day at daily intervals thereafter. The progression of the development of abnormalities is shown.

Toxicity in animals.—Single dose intraperitoneally: Mice: LD_{50} 280 mg/kg. Rats: LD_{50} 265 mg/kg. There is ataxia, weakness, and dyspnea within a few hours and death within a few days. Repeated dose toxicity: Mice: LD_{50} 100 mg/kg IP $\times$ 5. Rats: LD_{50} 80 mg/kg IP $\times$ 5. Dogs: LD_{50} 10 to 25 mg/kg IP $\times$ 10. There is seen diarrhea, jaundice, leucopenia, bromosulfalein retention. Also seen is intestinal congestion, swelling, and atypia of grandular nuclei and denuded epithelium. The marrow is severely hypoplastic. In a series of comparative toxic effects for rats and mice, the increasing order is purine, 6-chloropurine, 6-mercaptopurine, 2'6-diaminopurine, thioguanine, and 9-E-6-MP.

183   *Time and Dose-Response Relations*

Tumor inhibitory effects.—Mouse and rat transplanted tumors are inhibited by doses of 6-mercaptopurine of 50 mg/kg $\times$ 6 to 13.

Teratogenic effects.—The purine analogues inhibit the growth of the frog and rat embryos.

Clinical trials.—In man, doses of 2.5 mg/kg *per os* per day produce temporary remissions in acute leukemia and chronic myelogenous leukemia in children and adults. The drug has ben tolerated for several years in some cases. Higher doses will cause stomatitis, intestinal ulceration, and marrow depression. Samples of dosages and activity of purines: thioguanine at the same dose, 6-chloropurine at 20 mg/kg/day, thioguanine riboside and 6-mercaptopurine riboside at 2 mg/kg/day or 9-butyl-6-mercaptopurine at 10 mg/kg/day. Purine causes vesiculated skin lesions, 9-butyl 6-mercaptopurine and 6-chloropurine riboside have not been active.

Fetus from Each of Six Control and Six 6-Mercaptopurine-treated Litters (from Formalin-fixed Fetuses), Plate 8

Pregnant animals received 62.5 mg/kg of 6-mercaptopurine intraperitoneally on the twelfth day of gestation. Four days later, on the sixteenth day of gestation, a treated animal was sacrificed as well as a control for examination of fetuses. This continued each day thereafter until 21 animals were sacrificed. On the twelfth day, fetuses are extremely small and can be examined only with a dissecting microscope. However, by the sixteenth day, they are visible grossly. The first ossification centers are appearing. As development progresses, the shortness of the body and abnormalities of the limbs of the treated animals, as compared with controls, are demonstrated.

Acknowledgments

Table 1 and the first portion of Table 2 are from Murphy, M. L. Teratogenic effects in rats of growth-inhibiting chemicals, including studies on thalidomide. *Clinical Proceedings* (Children's Hospital of Washington, D.C.) **18** (11): 307, 1962.

Plates 3–5 are from Chaube, S. and Murphy, M. L. Teratogenic effect of Hadacidin (a new growth inhibitory chemical) on the rat fetus. *J. Exp. Zool.* **152**: 67, 1963.

Selected Bibliography

6-Mercaptopurine, edited by R. W. Miner. *Ann. N.Y. Acad. Sci.* **60**: 185, 1954.

Murphy, M. L. and Chaube, S. Teratogenic effects of abnormal purines and their ribosides in the rat. *Proc. Amer. Assoc. Cancer Research.* **3**(4): 347, 1962.

Data summaries of selected purine antagonists. Cancer Chemotherapy National Service Center. *Ca. Chemo. Reports* **11**: 117, 1961.

CHAPTER 8

MECHANISMS OF ACTION OF CERTAIN GROWTH-INHIBITING DRUGS

DAVID A. KARNOFSKY

Teratology, or the study of monsters, is not a defined scientific discipline. This book may help to form a conception of teratology as it relates to the origin of malformations in man. The relevance of a particular presentation to the subject of teratology or to the presumed purposes of this book may not appear to be direct and obvious, but each researcher provides material from which impressions can be derived, conclusions drawn, and work planned.

It is proposed as a law, which cannot be disproved, that any drug administered at the proper dosage, at the proper stage of development to embryos of the proper species—and these include both vertebrates and invertebrates—will be effective in causing disturbances in embryonic development. If we accept the statement that the ability to produce developmental abnormalities is a property of all drugs, our purpose is then to decide the laboratory circumstances under which a drug, in exhibiting significant teratogenic activity, might be expected to have similar effects on the human fetus. To the laboratory investigator, what is important is determined by his special interests. If he is interested in the frog embryo, a drug which modifies its development, such as benzimidazole, may fascinate him even though it may not produce similar effects in other species. Other investigators may be interested in a drug which affects the embryos of fish, *Drosophila*, salamanders, sand dollars, opossum, rats, or mice. Once the experimental system is defined, the methods of drug administration and the necessary observations are determined by the embryologist, the pharmacologist, the biochemist, the anatomist, or the geneticist concerned with the study. The work on developmental abnormalities is an end in itself, and an impulse for a particular study is not necessarily stimulated by a socially meaningful objective, such as using or studying an embryological system in order to detect the teratogenic potential of a drug for man.

The thalidomide disaster, out of necessity, gave our variously oriented

Sloan-Kettering Institute for Cancer Research, 444 E. 68th Street, New York, N.Y.

This research was supported, in part, by grants from the American Cancer Society, Inc. (T-40), The Albert and Mary Lasker Foundation, and grants from the National Institutes of Health (CA 03192-07S1 CY).

teratologists responsibilities that they were not prepared to accept, and may not even have been interested in accepting; that is, to try to prevent similar tragedies by developing sound laboratory criteria by means of which one could predict that a drug, given to a variety of human genotypes under different environmental conditions, would or would not have a significant teratogenic effect on the fetus. The experimental biologist, using the embryos of a specific species as a study system, is usually intent on producing developmental abnormalities, and he will alter his experimental conditions in the hope of accomplishing this objective. Society, the government, and the pharmaceutical industry have entirely different objectives, although they are forced to work with the same tools. They obviously hope to demonstrate by means of certain tests, the drugs which, when given at reasonable doses, will not have a teratogenic potential in pregnant women. This objective poses different problems and requires a different philosophy. While the techniques and experience of the experimental teratologists may be germane, their applicability to the problem of drug safety for the human embryo has not been defined.

Without pretending to be able to solve this crucial problem, my experience in cancer chemotherapy may be helpful in our analysis. My introduction to teratology stemmed from our work on the laboratory and clinical aspects of cancer chemotherapy. The functioning of this program can be discussed briefly and compared with the situation in laboratory and clinical teratology.

The clinical objective of the cancer program is to find drugs useful against the various forms of cancer in man. Since practically nothing is known about unique biochemical properties or carcinogenic factors operative within cancer cells, the chemotherapy program is almost entirely empirical—the indiscriminate testing of drugs in order to discover those with anticancer activity. The screening of drugs cannot be conducted initially in patients; the many candidate chemicals and antibiotic preparations must be screened first in laboratory animals bearing tumors—transplanted or spontaneous. The compounds showing activity in these systems are selected for clinical trial. Over the years an elaborate system has been devised to provide a flow of potentially active compounds for clinical study.

The steps in this screening program have been reviewed on several occasions (1–5). The described sequences of events are:

1. Drug procurement.—Procurement of chemicals, filtrates, biologicals, and tissue extracts from various companies. These preparations must have a known and reproducible origin, so that additional supplies can be obtained if an active compound is found.

2. Screening.—The preparations are screened against several forms of transplantable mouse tumors, the standard ones being sarcoma 180, carcinoma 755, and leukemia L1210. A number of other systems have been tried, including spontaneous and carcinogen-induced tumors, a variety of transplantable mouse tumors, ascites tumors, and transplantable tumors in rats, rabbits, and chickens. In other tests, human tumors have been grown in hamster, rat, and chick embryos and drugs given in order to inhibit their growth. Drugs are also tested against tumor cells in tissue culture and in various microbiological systems to find drugs with certain mechanisms of action. By this means, compounds producing evidence of growth inhibition and antitumor activity are selected for clinical study. It is to be understood that a number of compounds active against mouse tumors are inactive in man, and that some compounds have been found active in man before antitumor activity was demonstrated in laboratory animals.

3. Pharmacology.—The drugs selected for anticancer activity are studied by the biochemist to define their mechanism of action, and then by the pharmacologist in order to make recommendations concerning their dosage in man and to provide warnings concerning their toxicity at therapeutic doses. It is recognized that for the majority of anticancer drugs, there is a narrow therapeutic index and it is necessary to approach a near-toxic dose in order to give the patient the best chance for a therapeutic response.

4. Clinical trial.—After the pharmacological properties of the drug are defined, it is tested against various forms of cancer in man. At first, only patients with the far-advanced disease, not susceptible to any known form of treatment, are given the new drug. But as toxicological information is developed and evidence of therapeutic responses appear, less advanced cases can be treated with more certain therapeutic intent. The drug is finally evaluated under practical clinical conditions in a large number of patients, by comparison with other forms of treatment.

This method has turned up a number of drugs—many of them members of related groups—known to have some anticancer activity in man. While the practical benefit to patients from cancer chemotherapy has not been brilliant, a number of drugs have been temporarily useful against certain forms of cancer (6).

The empirical studies are continuing, and aspects of the cancer chemotherapy program are being reviewed and made more precise in order to obtain more reliable and quantitative data by determining the quantitative anticancer activity of a drug against a number of different kinds of cancers in animals, by studying the pharmacology of a drug of potential clinical interest in mice, rats, rabbits, cats, dogs, and monkeys, as well as by evaluating drugs in more detail in various forms of cancer in man.

We cannot reliably interpret, either quantitatively or qualitatively, results of animal screening tests in relation to the activity of the drugs against human cancer. This correlation cannot be established until we discover drugs which are selectively active against cancer in man. We can then use the data on the specific human anticancer compounds to measure the reliability and sensitivity of various animal test systems now being used to search for anticancer activity. Validation of the most reliable method in animals for detecting drugs of value for man would greatly simplify the screening program. For example, quantitative testing of compounds related to weak anticancer drugs against various forms of animal tumors has not produced derivatives with significantly enhanced effectiveness in man. The groups of anticancer compounds, e.g., the antifolics and alkylating agents, have been widely studied. Analogues within each group may show significant quantitative differences in activity against certain animal tumors, but active derivatives within each group have an essentially similar spectrum of anticancer activity in man. Screening has also selected a number of compounds active against certain animal tumors, but these drugs, thus far, have not shown appreciable therapeutic activity when tested against various forms of human cancer. These therapeutically ineffective compounds against human cancer include 2-ethylamino-1,3,4-thiadiazole, 6-aminonicotinamide, certain terephthalanalide derivatives, and cycloheximide.

Drugs with certain types of biological activity have been shown to be of value in the treatment of cancer. These drugs may be classified as follows:

1. Drugs which generally have an inhibitory action on cellular proliferation. These drugs damage both normal and neoplastic tissues which are undergoing frequent cell division. The normal structures most frequently affected in man are the bone marrow, epithelium of the digestive tract, germinal epithelium, hair, and lymphatic tissue. The active compounds include the following classes:

a. POLYFUNCTIONAL ALKYLATING AGENTS, such as nitrogen mustard, chlorambucil, triethylenethiophosphoramide (thio-TEPA), and busulfan.

b. ANTIMETABOLITES, which interfere with purine and pyrimidine biosynthesis and interconversion of purine precursors; these include the folic acid antagonists (methotrexate), azaserine, and 6-diazo-5-oxo-L-norleucine (DON), 6-mercaptopurine, and the 5-fluorinated pyrimidines.

c. METAPHASE INHIBITORS, such as colchicine, vinblastine, and vincristine.

d. ANTIBIOTICS, with a markedly cytotoxic activity such as actinomycin D, mitomycin C, and streptonigrin.

2. Steroid hormones, which through their physiological action influence neoplasms arising from tissues normally responsive to hormones. Cancers of the prostate, breast, uterus, and dyscrasias of the lymphatic system, such as acute leukemia, chronic lymphatic leukemia, and the lymphomas are responsive to the appropriate steroids.

3. Drugs acting on specific cancers because of their exaggerated functional properties—thus *o'p*-DDD (1,1-dichloro-2-(*o*-chlorophenyl)-2-(*p*-chlorophe-

nyl)-ethane) which blocks adrenal cortical hormone synthesis, influences the growth and function of functional carcinomas of the adrenal cortex.

At the moment, one can support the position that by selecting drugs which consistently depress the bone marrow of laboratory animals, hormones which influence the growth of specific tissues, and drugs which interfere with the growth or functions of certain normal cell types, all of the currently useful anticancer drugs might have been discovered without the use of tumor-screening methods.

This does not mean that animal screening methods should be abandoned. We are still looking for the anticancer drug with far greater selectivity which will destroy or control the growth of all cancer cells without damaging normal tissues or their functions. Until better methods are discovered to delineate the selective anticancer activity of a drug, the use of empirical screening methods must be continued.

The problems associated with laboratory analysis of drugs for teratogenic activity and the applicability of these results to man is comparable in some respects to the situation in cancer chemotherapy. These may be discussed in the same sequence as the teratogenic drugs.

1. Drug procurement.—All drugs should be tested in a suitable system for teratogenic activity before clinical trial. It would be wise to test a large number of drugs empirically in various systems in order to evaluate the significance of a teratogenic response under given circumstances.

2. Screening.—It is essential to develop acceptable screening methods to identify those drugs potentially teratogenic in man. The laboratory animals commonly used are the chick embryo, the pregnant mouse, rat, or rabbit. What is the proper way to test drugs in these animals in order to elicit evidence of specific injury or lack of injury to the fetus? Which is the most susceptible species? The embryos of which species most clearly resemble human embryos in their response to drugs? Under what circumstances are the results obtained in a test system applicable to man? Are there other test animals that may be more useful such as the armadillo, the frog egg, or the pregnant dog?

The value of a specific screening method is difficult to prove at present, and several must be tried until the most reliable one is found. In fact, until the different types of drugs teratogenic in man are known—and this possibility is hopefully remote—how can one decide which screening systems will supply results applicable to the clinical situation?

3. Pharmacology.—As more is learned about the mechanism of action of a drug and about the physiology of the fetus, it may be possible to anticipate the teratogenic action of certain compounds. This is an important research area. As the factors responsible for teratogenic activity are defined, those drugs exhibiting certain pharmacologic properties may demand the most careful laboratory scrutiny and clinical follow-up before they can be certified as safe for the pregnant woman.

4. Clinical trial.—It is not generally feasible to deliberately evaluate a drug for teratogenic activity in man. Data are being obtained from unplanned exposures of pregnant women to drugs (7), and in a unique study in pregnant women preliminary to therapeutic abortions, Thiersch (8) showed that aminopterin was teratogenic. In suitable circumstances, it seems reasonable to design clinical trials to confirm an impression that certain drugs are *not* teratogenic at reasonable doses.

In contrast, cancer chemotherapy is in a strong position to correlate laboratory with clinical results. Drugs can be evaluated for therapeutic activity at maximum tolerated doses against various forms of human cancer; this is done even if a drug possesses considerable toxicity, because nothing else of value may be available in a desperate situation.

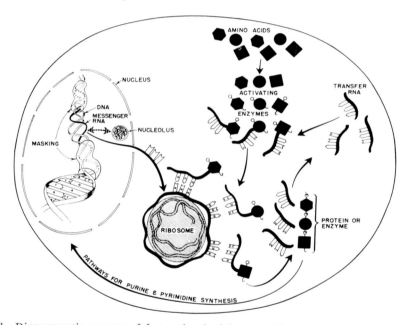

Fig. 1.—Diagrammatic concept of factors involved in controlling synthesis of specific proteins. (From [9].) m-RNA is formed through the action of an RNA polymerase on a "structural" component of the DNA which has a complementary base sequence. The role of histones and other chromosomal proteins in "masking" portions of the DNA not used in templating m-RNA, the mode of action of "regulatory" genes and repressor and inducer substances that govern the rate of m-RNA synthesis, and the function of the nucleolus are not clear at the present. The m-RNA passes into the cytoplasm where in some manner it attaches to a ribosome to direct specific protein synthesis. The function of the pre-existent "workshop" ribosomal RNA and its relationship with m-RNA are not known. Amino acids are activated by specific enzymes and transported to the ribosomes by specific t-RNA molecules. These specific t-RNA molecules are aligned at particular sites on the m-RNA through possession of a complementary base sequence, and thereby assure transcription of the proper peptide bonding necessary for assembly of the amino acids into specific enzymes. These enzymes act on specific substrates along defined biosynthetic pathways to produce the more complex materials necessary for cell function and reproduction.

David A. Karnofsky   190

In the absence of reliable clinical testing facilities for teratogenic activity, the major permissible teratological programs lie in biochemical studies on embryonic development, in drug screening with various species of pregnant animals, and in studies on the mechanisms of drug action. It is in the latter area that the work of investigators in chemotherapy and teratology have converged.

Many individuals engaged in screening drugs in cancer chemotherapy have diverted their interest toward the mechanism of action of the various anticancer drugs in inhibiting cell growth, both normal and neoplastic. Some of the most brilliant recent work in biochemistry and molecular biology has resulted from studies on drugs used in cancer chemotherapy (9). From our knowledge of the mechanism of action it was assumed that anticancer drugs would very likely be teratogenic in various embryonic systems, including the human fetus. Our predictions were well-supported in laboratory animals, and they probably would be confirmed in man if the necessary studies were feasible.

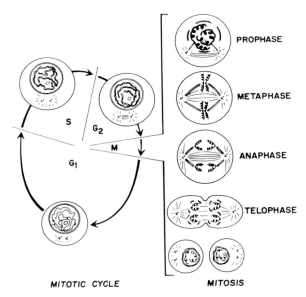

Fig. 2.—Diagrammatic representation of major events taking place during the different phases of the division cycle:

G$_1$ (postmitotic interval) is the interval following mitosis and before the beginning of DNA synthesis, and usually comprises over half the total generation time.

S is the period of DNA synthesis; this comprises about one-fourth to one-third of the generation time.

G$_2$ (premitotic interval) is the period between the end of DNA synthesis and the beginning of mitosis, and comprises less than one-fifth of the generation time.

M is the period of mitosis.

The generation time is the time it takes for a cell to complete one entire division cycle. (From [9].)

Growth-inhibiting Drugs

Current views of cell activities during proliferative periods and of mechanisms of the action of drugs which inhibit growth are shown in the following figures.

Figure 1 illustrates the structural components of a proliferating cell and the biosynthetic events related to cell growth.

Figure 2 shows the stages in cellular growth and mitosis.

Figure 3 lists anticancer drugs, most of which have been shown to be toxic to the rat fetus, and in many cases, to produce developmental abnormalities. Their primary sites of action are noted as (1) on DNA, either in altering its structure in being directly incorporated into DNA, or in blocking DNA dependent RNA systems; (2) on RNA, in producing abnormal purines or pyrimidines containing RNA; (3) on protein synthesis; and (4) on nucleic acid synthesis, either through blocking purine or pyrimidine biosynthesis or interconversion of purines.

Figure 4 lists drugs with other mechanisms of action, namely steroids and plant alkaloids.

It is reasonable to suggest that other potent growth-inhibiting drugs may be expected to be teratogenic in mammalian systems.

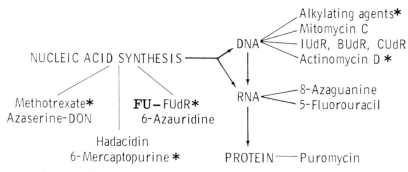

FIG. 3.—Sites of action of anticancer drugs on major components and metabolic pathways of cells. DNA (deoxyribonucleic acid) is altered in various ways by alkylating agents such as nitrogen mustard; mitomycin C, an antibiotic; certain 5-halogenated pyrimidines, iodo-(IUdR), chloro-(CUdR), and bromodeoxyuridine (BUdR); and actinomycin D, an antibiotic. Actinomycin D complexes with DNA and blocks the formation of RNA from the DNA template.

RNA (ribonucleic acid) is modified by the introduction of an abnormal purine or pyrimidine. Two compounds that have been shown to do this are 8-azaguanine and 5-fluorouracil. In damaging cells *in vivo*, however, 5-fluorouracil (FU) acts principally by being converted to 5-fluorodeoxyuridine (FUdR), which inhibits thymidylate synthetase and thus prevents the methylation of deoxyuridylic acid to form thymidylic acid. This produces a thymidine deficiency and thus inhibits DNA synthesis.

Protein synthesis occurs on the ribosomes. Puromycin has been shown to interfere with protein formation by the ribosomes.

Another major pathway is the biosynthesis of nucleic acids for formation of DNA, RNA, and acid-soluble nucleic acids within cells. Methotrexate and azaserine inhibit the biosynthesis of purines. Hadacidin and 6-mercaptopurine interfere with the interconversion of purines necessary for cell growth. FUdR inhibits thymidylate synthetase, and 6-azauridine blocks the decarboxylation of orotic acid to form uracil. The asterisk (*) refers to those compounds that have been of particular value in the treatment of cancer in man.

The mechanisms shown in Figures 3 and 4 and the drugs acting by these mechanisms are not the only means whereby drugs cause developmental abnormalities. They are merely illustrative of an important segment of the problem which has been worked out in some detail and has some relevance to the human situation. Other substances, however, such as some vitamins and vitamin antagonists, acting by other mechanisms, are teratogenic. In animals there are a number of teratogenic drugs whose mechanism of action has not been defined at all. We must construct systems in which drug effects on cells and tissues can be correlated with patterns in teratologic responses.

While we discuss the need for basic studies in teratologic pharmacology,

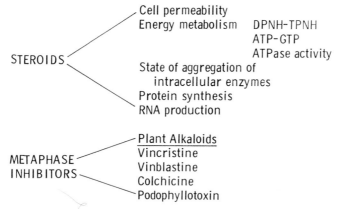

FIG. 4.—Steroid hormones and plant alkaloids which act on cell growth.

The steroid hormones act on specific responsive tissues to either inhibit or to stimulate their growth. The site of action of steroid hormones has been the subject of many studies and the various proposed sites of action are noted.

The plant alkaloids have a common effect in the inhibiting cells in metaphase. There are certain qualitative differences, however, in their therapeutic effect, in that vincristine appears to be more specific than vinblastine in leukemia and colchicine appears to be the most effective member of the group for the treatment of gout.

the bugaboo of thalidomide continues to plague us and increases the urgency for action. No matter how carefully we devise a test system, if the teratogenic effect of thalidomide cannot be detected at reasonable doses and its action on the human fetus explained, we are obviously vulnerable to the possibility that other drugs, acting insidiously, may slip through screens designed to detect teratogenic activity. No one in the pharmaceutical field can rest easily until the thalidomide problem is solved. This is the first order of business.

REFERENCES

1. STOCK, C. C. Experimental cancer chemotherapy. *Adv. Canc. Research* **2**: 425, 1954.
2. FARBER, S., TOCH, R., SEARS, E. M., AND PINKEL, D. Advances in chemotherapy of cancer in man. *Adv. Canc. Research* **4**: 1, 1956.
3. Conference on Cancer Chemotherapy Screening Procedures. *Annals N. Y. Acad. Sci.* **76**: 409, 1958.

4. Cancer Chemotherapy Reports, published under the auspices of the Cancer Chemotherapy National Service Center.
5. LEITER, J., ABBOTT, B. J., SCHEPARTZ, S. A., AND WODINSKY, I. Screening data from the cancer chemotherapy national service center screening laboratories. *Cancer Research Supplement* **24**(part 2): 473, 1964.
6. KARNOFSKY, D. A. Cancer chemotherapy agents. *Ca—a Cancer Journal for Clinicians* **14**: 67, 1964.
7. SOKAL, J. E. AND LESSMAN, E. M. Effects of cancer chemotherapeutic agents on the human fetus. *JAMA* **172**: 1765, 1960.
8. THIERSCH, J. B. Therapeutic abortions with folic acid antagonist, 4-aminopteroylglutamic acid (4-amino PGA) administered by the oral route. *Amer. J. Obst. Gynec.* **63**: 1298, 1952.
9. KARNOFSKY, D. A. AND CLARKSON, B. D. Cellular effects of anticancer drugs. *Annual Rev. of Pharmacology* **3**: 357, 1963.

THE CHICK EMBRYO IN DRUG SCREENING; SURVEY OF TERATOLOGICAL EFFECTS OBSERVED IN THE 4-DAY CHICK EMBRYO

The developing chick embryo is one of the most extensively used living systems for biological research. The availability of fertile eggs, the rapid growth of the embryo, and the ease in manipulating it and altering its environment have made the chick embryo a model system for morphological, biochemical, and functional studies on growth, differentiation, and organogenesis. During the 21-day period when the egg progresses from a single cell to a hatched and self-sufficient individual are concentrated most of the complex problems of development and differentiation. Detailed descriptions of these events are provided by several monographs and textbooks (1–7).

Many drugs have been studied for their effects on the developing chick embryo (6, 8). Our researchers became interested in the effects of drugs on embryonic development during the course of testing various compounds empirically for specific inhibitory effects on mammalian tumors growing on the chorioallantoic membrane of the chick embryo (9). Some of the compounds, particularly growth-modifiers, produced developmental abnormalities. As a result, the developing chick embryo became the basis of an independent method for screening compounds for teratogenic activity (10).

GENERAL CONSIDERATIONS

The following arbitrary procedures, to some extent based on experience, are used in order to standardize and facilitate the screening of test compounds in the chick embryo. If the importance of a test compound is indi-

This article is adapted from the *Stanford Medical Bulletin*, Vol. 13, 1955.

David A. Karnofsky   194

cated by other studies, or if preliminary results in the chick embryo are of interest, a more detailed analysis of the activity of the agent is undertaken.

Chick embryos.—Under ideal conditions, it is desirable to use embryos which are both genetically and nutritionally uniform. Some strains produce a large number of abnormal embryos, and there are strain differences in the embryonic response to certain drugs (11). The maternal diet affects the nutrient and vitamin content of the egg (1), and certain noxious substances may actually be introduced from the mother's diet into the egg (12). These factors can alter the results or confuse their interpretation. In our studies White Leghorn eggs, obtained from a commercial source, are used. Our eggs have been supplied by the Shamrock Poultry Farm, South Boyd Parkway, North Brunswick, N.J.

Incubation.—A Baby Mammoth Electric Incubator, American Lincoln Incubator Company, New Brunswick, N.J., has been used. The embryos are incubated at 38° C and a relative humidity of 65 to 85 per cent. The eggs are candled daily for viability. The candler is the Prevue Fertility Tester, Breeder's Supply Company, New York, N.Y. After the tenth day of incubation, if candling reveals a restricted chorioallantoic membrane, it is likely that the embryo is stunted and abnormal. The dead embryos are inspected grossly. While abnormal, dead embryos cannot be classified satisfactorily, unusual and sometimes exaggerated alterations may be seen that can then be sought for more carefully in the surviving and sacrificed embryos.

Age of embryo at injection.—The age of the embryo at the time of treatment is of critical importance. The embryo is constantly evolving and changing, with new systems and functions appearing and disappearing. Given too early in development, a specific drug may have no sensitive system on which to act; given too late, the sensitive system may have disappeared, or protective systems may have developed which detoxify the drug or actively compete for it with the sensitive system. Also, a large dose of a drug, tolerated early in development because a vital, sensitive system has not yet appeared, may be able to produce morphological abnormalities. If given at later stages, when a vital system sensitive to the drug has appeared, the drug may have an acute lethal action at small doses and the embryo will not tolerate a dose sufficient to produce morphological changes in the survivors.

The most striking and consistent developmental abnormalities are usually produced when the embryo is treated around the fourth day of incubation. Earlier, the mass of embryo is small, and death may be due to non-specific action of the relatively high concentration of the drug or the embryo may be so sensitive that only subtle changes are produced during the short period of survival following treatment. When organogenesis is well established, however, test compounds can selectively affect certain systems and organs without necessarily seriously intoxicating the entire embryo, and thus, very specific developmental abnormalities will be produced. While the 4-day period is the most satisfactory one in general for demonstrating specific effects, some agents exert a characteristic action on the embryo when given over the range of 0 to 8 eight days of incubation or later. For most studies we have injected the test compounds on the fourth day of incubation.

Preparation of solutions.—All solutions are prepared under sterile conditions. Soluble materials are made up in physiological saline. If the drug is not soluble at the necessary concentrations, it is suspended in a 0.5 per cent solution of carboxymethylcellulose (CMC) in saline. Some highly insoluble materials, however, when given as suspensions, are apparently not readily absorbed from the yolk. Whenever possible, it is a good rule to convert the drug to a soluble form by forming a soluble salt or by altering the pH of the solution. It must be shown, however, that these procedures do not change the structure of the drug. In some instances the drug can be given in small doses of an organic solvent.

Methods of injection.—Injections can be made into the yolk sac, allantoic sac, intra-embryonically, intravenously, or onto the chorioallantoic membrane (CAM). The simplest and usually the most satisfactory method is yolk sac injection; the volume usually ranges from 0.1 to 0.2 cc, although larger volumes of fluid are tolerated. Some drugs are more active or exhibit different specific effects if other routes of injection are used. Thus cortisone acetate is three to four times more active in inhibiting embryonic growth by CAM as compared to yolk sac injection (13). Ferrous sulfate is acutely toxic when injected on the CAM, but much larger doses, which sometimes produce developmental abnormalities, can be tolerated by the yolk sac route, possibly because an acute lethal level is avoided by the slower yolk absorption (14). Single doses are usually given. The egg is a closed system. Injected drugs are usually dispersed, absorbed, and in various ways distributed throughout the egg. Single doses are probably satisfactory in most cases; in rare instances it is possible that multiple small doses may result in developmental abnormalities, whereas a large single dose would be acutely lethal or fail to produce morphologic effects.

Dosage.—In preliminary studies, by assuming adequate amounts of the test drug are available, doses of 0.1, 1.0, and 10.0 mg/egg are given to four eggs each. On the basis of the survival of the embryos and evidences of developmental abnormalities, a second dosage schedule is run in the tolerated dose range; e.g., if all embryos survived at 0.1 mg, and all died at 1.0 and 10.0 mg/egg, the dosages for the second run would be 0.2, 0.4, and 0.8 mg/egg. Usually six eggs are injected at each level. The results of the second run will usually decide whether further work on the candidate drug is indicated.

Mortality and survival.—In general, active drugs produce the most exaggerated developmental abnormalities when given in the range of the maximum tolerated dose. Surviving embryos are sacrificed usually at 16 to 18 days of incubation; at this time the form of the embryo resembles that at hatching and gross abnormalities are most readily identified. Some compounds, however, may cause a peak of deaths at the eleventh to twelfth days, or fourteenth to sixteenth days, and embryos dying or sacrificed at this time show severe and characteristic abnormalities. The embryos surviving beyond the critical period, however, may be much less severely affected or may appear normal. Thus the selection of an arbitrary day of sacrifice, such as sixteen or eighteen days, without regard to the distribution of deaths produced by the drug, may result in the omission of many interesting observations. The toxicity of a drug is usually recorded as the LD_{50} dose (dose lethal to 50 per cent of the embryos within 10 days after injection) in relation to a given embryonic age and route of injection, and the distribution of deaths is described in the range of the LD_{10} to LD_{90}.

Cause of death.—The cause of death in the embryo is usually difficult to determine. Some embryos die at a specific time after injection with no apparent morphological disturbance (15), others die at critical periods with severe developmental defects, and others, while grossly abnormal, may survive until the time of hatching but are unable to hatch. It is of interest that embryos almost totally decerebrated (16) or with severe brain hemorrhages (17) often survive until the time of hatching.

Summary of techniques for injecting the chick embryo.—4-DAY YOLK SAC INJECTION. The eggs are candled for viability, and the blunt end wiped with 70 per cent ethyl alcohol. An egg punch (Virtus Co., Yonkers, N.Y.) is used to punch a small hole in the blunt end of the egg. The solution is injected through the hole in the blunt end into the yolk sac by means of a tuberculin syringe with a 1.5-inch 21–24 gauge needle. The hole is then sealed with paraffin.

8-DAY CAM INJECTION. The eggs are candled for viability, and the area containing the major blood vessels is circled. This area is cleaned with 70 per cent ethyl alcohol. The egg punch is used to punch a hole in the blunt end. A drill with an emery disc is used to cut a small square opening over the blood vessels. The shell is lifted off with sterile forceps. The shell membrane is dampened with isotonic saline, and the membrane is removed with forceps. The CAM will drop. The solution containing the drug is then dropped onto the CAM. The opening is then sealed with cellophane tape.

Developmental abnormality classification.—When the embryos are sacrificed, they are inspected carefully for evidence of developmental defects. There are numerous abnormalities—genetic, spontaneous, and induced— which have been described and classified by teratologists. The age of the embryo and the methods of observation used—gross inspection, histological studies, reconstructions, histochemical and biochemical determinations, functional studies, etc.—largely determine the type of information obtained. It is desirable, of course, to examine the embryo by all these techniques, so that the whole story will be elucidated and nothing of importance will be missed. This is obviously not feasible. In a screening procedure simplicity is essential, and thus only certain gross observations are routinely made. Embryos are sacrificed after the twelfth day, preferably on the sixteenth to eighteenth days of incubation. Routine observations include:

INSPECTION FOR OVER-ALL APPEARANCE. The following disturbances may be seen: stunting in size, edema, corneal cysts, eyelid defects, exteriorization of viscera, thin legs, defects in feather development, facial coloboma, cleft palate, parrot beak, micromelia, missing or webbed toes, suggestion of fused vertebrae, brain hemorrhage, and weight of embryos.

EXAMINATION OF SYSTEMS.

1. Skeleton. While some skeletal defects may be found on gross inspection, the skeletons of selected embryos should be cleared and stained with alizarin red S. For clearing, the abdominal cavity is opened and the embryo is fixed in 95 per cent ethyl alcohol. After 3 days, the embryo is eviscerated. It is left in fresh 95 per cent alcohol for 1 week and then transferred to acetone for 2 weeks.

After the exposure to acetone, it is returned to 95 per cent ethyl alcohol for 2 days. The embryo is then placed in 1 per cent potassium hydroxide until the tissues are clear and the bones opaque. About 24 hours after being placed in KOH the embryos are removed briefly and defeathered.

The cleared embryo is then placed in 0.5 per cent KOH, and 3 drops of a saturated aqueous solution of alizarin red S is added. The embryo is removed after approximately 16 hours. The embryo is then subjected to glycerine in increasing concentrations, 30, 60, 90, and 100 per cent, remaining in each concentration except the last for 48 hours. The embryo can be kept in pure glycerine to which a few crystals of thymol have been added. The embryos are inspected, and the osseous structures described.

2. Brain. Meningoceles and hemorrhage may be seen externally: the skulls are split and the brains examined grossly.

3. Feathers. It is time consuming to examine feather tracts, but feather growth may be markedly retarded or feathers may be clubbed, pale, and fine. Pigment changes cannot be studied in the White Leghorn, of course, but they are of considerable interest in other strains (18).

4. Eyes. These may be small, have corneal cysts, and occasionally a defect in the upper lid.

5. Liver and spleen. The liver may appear small, with necrotic patches

on the edge, or enlarged, pale, and fatty. The spleen may be small or enlarged. These organs are weighed if they appear grossly abnormal.

Many more observations can be made, of a gross, histological, or biochemical nature. The embryo can be treated at various days of incubation, the drug can be given in combination with others, or protective substances can be added to specifically counteract the effect of a toxic compound. Labeled precursors are helpful in determining the mechanism of action of a compound. Once a drug is found to have a consistent effect on the embryo, a variety of studies on its mechanism of action is of interest.

In any work involving chick embryos, the occurrence of spontaneous

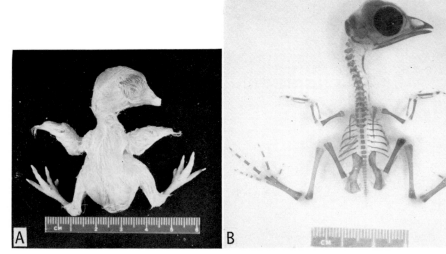

PLATE 1.—*A*, 18-day control embryo, and *B*, 18-day embryo cleared and stained with alizarin red S.

deaths and developmental defects must be evaluated. To be considered active in the chick embryo, any drug must produce a characteristic and consistent abnormality in an appreciable number of embryos on several independent trials, and the alterations in the embryo should not resemble, in type or frequency of occurence, those seen in the controls.

PATTERNS AND EFFECTS OBSERVED WITH REPRESENTATIVE
CLASSES OF DRUGS

Plate 1 shows a normal 18-day embryo and the stained skeleton of the cleared 18 day embryo. In subsequent plates some of the characteristic patterns of effects consistently produced by certain drugs on the 4- to 8-day chick embryo will be shown.

Micromelia.—One of the most common teratogenic syndromes in the chick embryo is a reduction in the size of the bones of the extremities; this is referred to as micromelia (Plate 2). While the chemicals that produce this effect differ considerably in structure and pharmacologic action, the embryo can be protected against a number of them by the injection of nicotinamide. Micromelia develops in embryos treated with appropriate drugs at 4 days, but the drugs produce similar effects when given up to the seventh and the eighth days of development. This suggests that these agents act on

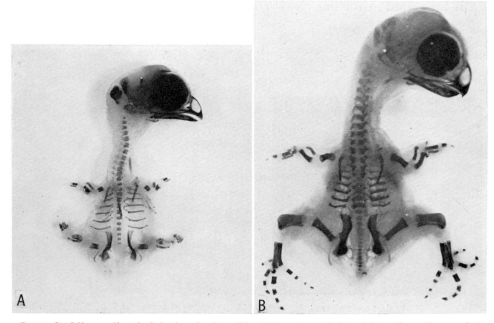

PLATE 2.—Micromelia. *A*, 6-Aminonicotinamide, 0.02 mg/egg, injected into the yolk sac of the 4-day embryo. Note the curved upper beak, the severe micromelia, and the poorly ossified toes. *B*, Methionine sulfoximine, 1 mg/egg, injected into the yolk sac of the 4-day embryo. Note the parrot upper beak, the severe micromelia, and the tibia slightly bent at the distal end.

systems in the developing skeleton which are sensitive up to the eighth day of incubation.

The drugs which cause micromelia and whose effects are prevented by the administration of nicotinamide include 6-aminonicotinamide (19–21) (Plate 2*A*), 5-fluoronicotinamide, pilocarpine hydrochloride (22), insulin (23), sulfanilamide (24), and a triazene derivative (3,3-dimethyl-l-phenoyl-triazene) (25). Methionine sulfoximine produces micromelia (Plate 2*B*, which cannot be prevented by nicotinamide (26). Thallium produces micromelia as well as a reduction in the size of all bones (27). Apparently, thallium acts directly on the cartilage and blocks the progressive growth of endochondral bone, although periosteal bone formation continues. The micromelia caused by these drugs may differ quantitatively, but there are general similarities. Micromelia is associated in most embryos with parrot beak (curved upper beak and slightly shortened lower beak).

Fusion of vertebrae.—Fusion of vertebrae, to varying degrees of severity, has been produced by certain quaternary ammonium compounds, such as tetramethyl ammonium bromide and decamethonium bromide, and by physostigmine (28). The vertebral fusion begins at the cephalad end and with increasing doses extends posteriorly. The severity of the fusion produced by neostigmine is shown in Plate 3; the other bones are relatively

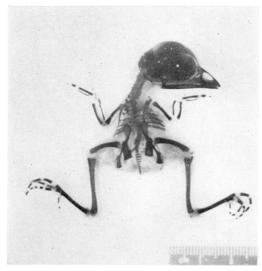

PLATE 3.—Fusion of vertebrae. Neostigmine methylsulfate, 1 mg/egg, injected into the yolk sac of the 4-day embryo. Note that the cervical vertebrae are fused, and the spinal column is twisted; the bones of the extremities appear to be normal, except for the angulation of the femur.

normal except for angulation of the femur. The specificity of the effect is illustrated by the fact that physostigmine, a potent cholinesterase inhibitor, will produce the fused vertebrae as well as the micromelia–parrot-beak syndrome, whereas pilocarpine (22), physostigmine, and neostigmine have a different range of effects on skeletal development.

Stunting and skeletal deformities produced by antimetabolites.—There are a number of metabolites and antimetabolites interfering with the purine and pyrimidine biosynthesis in metabolism (29). These drugs cause profound growth inhibition when injected into the yolk sac on the fourth day of incubation. Embryos may survive for 12 to 18 days days, but at sacrifice they are small, feather development is impaired, and the extent of malformations may be extraordinarily severe. The most active compounds in this group are the folic acid antagonists, azaserine, deoxyguanosine, 5-halogenated pyrimidines, and cytosine arabinoside. While their effects differ in some respects, the period of maximal embryonic sensitivity and the affected structures are closely related.

Azaserine (O-diazoacetyl-L-serine) blocks the step in the transfer of an amino group to formylglycine amide ribotide (FGAR) to form formylglycine amidine riboside (FGAM); thus it interferes with *de novo* purine synthesis. Embryos treated at 4 days are affected severely by azaserine and

PLATE 4.—The antimetabolite, azaserine, 0.15 mg/egg, was injected into the yolk sac of the 4-day embryo. Note the facial coloboma with cleft palate and shortened mandible, short-bent metatarsals, missing metacarpals, and missing toes and wing phalanges.

have beak abnormalities, facial coloboma with cleft palate, and appendicular defects of both the wings and the legs, including micromelia, hypophalangy, ectrodactyly, and hemimelia (30) (Plate 4). Azaserine also causes a tendency to micromelia and parrot beak when given at 8 days, but no skeletal malformations are seen in the embryos treated at 12 days. Adenine will prevent the action of azaserine on the skeleton of the chick embryo (31).

Developmental defects also are produced with the normally occurring purine, deoxyguanosine (dGuR) (32). Apparently, this substance acts by interfering with the formation of deoxycytidine (dCR) since the injection of dCR in the chick embryo will prevent the teratogenic effects of dGuR (33).

The severe effects of dGuR on the 4-day chick embryo are shown in Plates 5–8. Bone deletions and the absence of the knee and the elbow joints in many of the embryos treated with dGuR are of particular interest. While severe defects are produced when the drug is given at 4, 5, or 6 days, by the seventh day, dGuR is no longer a potent teratogenic agent.

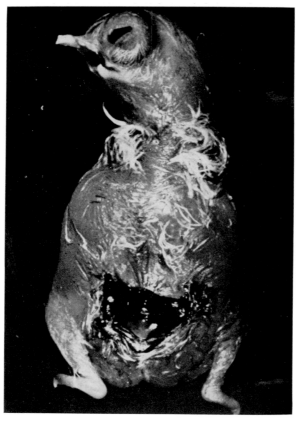

PLATE 5.—The antimetabolite, 2′ deoxyguanosine (dGuR), 0.5 mg/egg, was injected into the yolk sac of the 4-day embryo. Note facial coloboma, cleft palate, short lower beak, eyelid defect, edema, and feather inhibition, micromelia, and missing toes.

The pyrimidines, halogenated in the 5 position, interfere with pyrimidine metabolism. Fluorouracil (FU) may be metabolized to the ribosyl or deoxyribosyl form, to be incorporated into RNA or to form 5-fluorodeoxyuridine (FUDR) which blocks a step involving the methylation of deoxyuridylic acid (dUMP) to form thymidylic acid. FU will produce marked bony defects when injected in the 4-day embryo at a dose of 0.2 mg/egg (Plate 9). The toxic dose of FU is markedly decreased in the 7- to 8-day embryo, and its teratogenic activity is diminished. FUDR is far more active by weight,

Drug Screening

and 0.08 gamma/egg is in the lethal dose range for the 4-day embryo. The effects produced in the surviving embryo at LD_{50} doses are similar to those seen with 5-FU (34). Pyrimidines with other halogens in the 5 position (I, Cl, Br) appear to act by a different mechanism—either by competing with thymidine for a phosphorylating enzyme or by being directly incorporated in place of thymidine into DNA. These drugs do not produce many de-

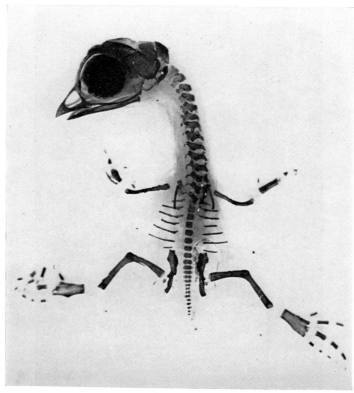

PLATE 6.—The antimetabolite, 2' deoxyguanosine (dGuR), 0.25 mg/egg, was injected into the yolk sac of the 4-day embryo. Note the fusion of the femur and the tibia, the fusion of the humerus and the ulna, and the fused short metatarsals and metacarpals.

fects in embryos surviving to 18 days, except for an interesting duplication of the first toe.

Growth-inhibition by adrenal steroids.—The adrenal streoids have specific and profound effects on the body growth and the osseous development of the chick embryo (13, 35). Embryos injected at 4 days are affected less severely than those injected by the CAM route on the eighth day of incubation, and the effects produced in the 4-day embryo differ qualitatively from those treated at 8 days. Plate 10*A* shows a stunted embryo treated at 8 days with an adrenal steroid derivative, which produces effects similar to cortisone on the chick embryo. The bones are all present in the cleared skeleton, but they stain faintly and are reduced in size. The 4-day embryo shows specific skeletal defects, whereas in the 8-day embryo the adrenal steroids apparently interfere mainly with the process of ossification since cartilaginous precursors appear to be normal.

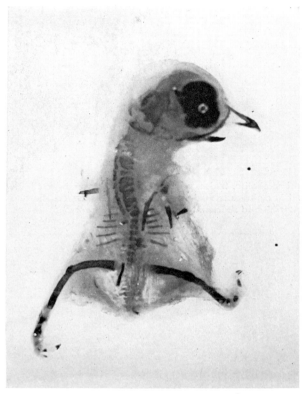

PLATE 7.—The antimetabolite, 2' deoxyguanosine (dGuR), 0.5 mg/egg, was injected into the yolk sac of the 4-day embryo. Note the facial coloboma, cleft palate, short lower beak, the fusion of the femur and the tibia, the missing radius, ulna, metacarpals, and wing phalanges, and the short and missing metatarsals and toes as well as the malformation of the vertebrae and hemimelia.

The adrenal steroids produce a varied pattern of exteriorization of the viscera, depending on the age of the embryo when treated. In contrast with the rather characteristic pattern of micromelia, which is not precisely age-related up to 8 days, the osseous effects and disturbances in feather pattern seen with the adrenal steroids vary considerably with the age of the embryo at the time when they are applied. The chick embryo has been useful in studying the structure-function relationships of the adrenal steroids. There is a marked specificity in the structural changes in the steroids and their

Drug Screening

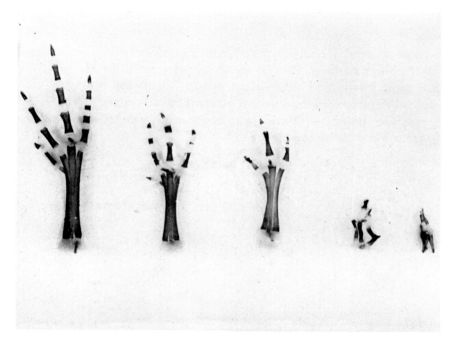

PLATE 8.—The antimetabolite, 2'-deoxyguanosine (dGuR), was injected. Note the range of abnormalities produced in embryos treated on the fifth day. Defects are shortening, fusion, bending, and absence of metatarsals and shortening and absence of the toes.

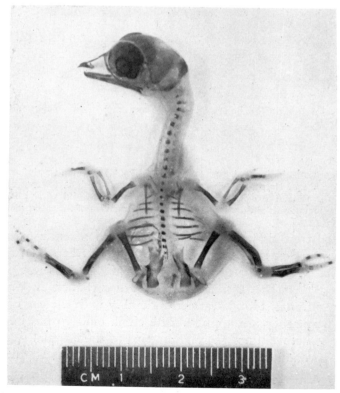

PLATE 9.—The antimetabolite, FU, 0.2 mg/egg, was injected into the yolk sac of the 4-day embryo. Note the facial coloboma and the cleft palate, the slight shortening of the lower beak, the bending of the tibia and metatarsals, and poor ossification in the bones of the toes.

effects on the chick embryo. A uniform pattern of biological effects is produced at doses ranging from 2 mg down to 0.025 gamma/egg depending on steroid administered.

Amino acid antimetabolites.—Ethionine is an antagonist of methionine. In the chick embryo, ethionine caused marked enlargement of the liver (Plate 10B) with increase in liver fat and thin legs, presumably due to im-

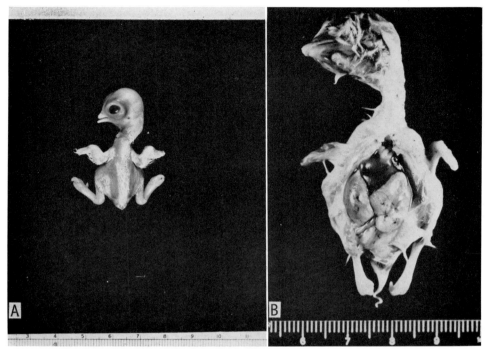

PLATE 10.—Adrenal steroids and amino acid antimetabolites. *A*, 21-Hydroxyoxylone, 0.1 gamma/egg inoculated on the CAM of the 8-day embryo. Note the short upper beak, the stunting, and the impaired feather development. *B*, DL-Ethionine, 8 mg/egg, was injected into the yolk sac of the 4-day embryo. Note the infiltration of fat into the liver.

paired muscle development. These embryos occasionally have a large rump sac. Methionine protected the embryos against the action of ethionine (36).

Drugs damaging the central nervous system.—Lead nitrate caused prompt and extensive brain hemorrhages when injected on the fourth day. Subsequently meningoceles appeared, although there was some recovery of brain substance; when lead is given on the eighth day, brain hemorrhage also occurs and the brain remains liquefied without restoration of brain substance (17). Cobalt chloride caused cerebral hemorrhage considerably different from the lead-induced injury in that it appeared late and was associated microscopically with a diffuse but less drastic hemorrhage and neuronal injury (14). Also, the liver was enlarged and fatty, and in some embryos the spleen was pale and enlarged; the latter was not due to hematopoietic activity but to a myxomatous stroma. The cobalt-treated embryos were also pale, and the severely affected ones usually died during the fourteenth to sixteenth days.

Discussion

The various organ systems of the chick embryo have been shown to respond, in a variety of ways, to the action of certain drugs. The alterations in embryonic development produced by a drug is evidence of specific biological activity, and in some cases the effects have suggested, by analogy with other drug effects, the mechanism of drug action. Alterations in embryonic development have been studied by gross inspection and by clearing the embryo for skeletal examination.

Stunting of growth.—This may occur without any other obvious abnormality; in fact, Zwilling and DeBell (24) have shown that nicotinamide will prevent the micromelia but not the growth inhibition that is induced by sulfanilamide. Stunting may be due to some direct effect on the embryo; but, also, a direct or indirect action on the extra-embryonic membranes with interference in growth, respiratory exchange, or absorption of yolk or albumen must be considered. Deficient membrane formation is seen following cortisone, the folic acid antagonists, and azaserine. In the cortisone-treated embryos, which may survive to hatching, large amounts of the yolk and albumen are unused. Consistent inhibition of embryonic growth, therefore, represents a highly significant drug action.

Skeletal abnormalities.—The skeletal system is easy to study in older embryos and, because of its many components, interesting patterns result from the actions of different drugs. Landauer (8) has examined the skeletal system intensively and described the patterns of skeletal disturbances produced by a number of drugs. He has shown that some agents, such as boric acid and pilocarpine, cause severe facial coloboma and micromelia with bone deletions, whereas other agents—physostigmine, sulfanilamide, and insulin—produce parrot beak and micromelia without bone loss. In our own work, azaserine tends to fall into the boric acid category, although there are definite differences (30), whereas methionine sulfoximine more closely resembles sulfanilamide and insulin. The compounds producing the facial coloboma and bone deletions (boric acid, azaserine, and pilocarpine) are most effective when given around the fourth day, and lose this activity as the embryo grows older, whereas insulin and sulfanilamide are active in producing micromelia and parrot beak when given as late as the eighth day of incubation. Of further interest, in indicating the specificity of these effects, is Landauer's demonstration that nicotinamide will protect against the micromelic effects of sulfonamides, physostigmine, and insulin, but exerts a weak action against the micromelic effects of pilocarpine (22). Thallium is effective when given as late as the eighth day. Although the centers of bone formation are present, thallium appears to interfere with the processes of bone growth and calcification.

The fusion of the vertebrae, in characteristic and different patterns, is produced by certain quaternary ammonium compounds and physostigmine. It is clear that this vertebral effect can be dissociated from the micromelic activity since pilocarpine produces the latter, tetramethylammonium bromide the vertebral fusion, and physostigmine both effects, although the latter is not identical with those produced by each of the other compounds alone. Cortisone appears to cause chronological retardation in osseous development.

Brain.—Some information can be gained from gross inspection of the central nervous system, but any fundamental analysis requires microscopic study. Cortisone and the folic acid antagonists apparently produce small brains, but whether there is uniform retardation in growth or inhibition of a selected portion of the brain is not known. Thallium also affects the brain, apparently by inhibiting the growth of the skull. Brain hemorrhages produced by lead have resulted in meningoceles. Cobalt produced a delayed brain hemorrhage without meningoceles.

Liver.—The liver begins to function about the eighth day, and subsequently gross changes in the liver become evident. Many compounds produce necrotic patches, particularly on the liver edges. Ethionine and cobalt cause liver enlargement due to infiltration with fat. Methionine will protect the liver against ethionine, but not against cobalt. Subtle derangements in liver function possibly may be responsible for other defects in embryonic development. Landauer (37), for example, has shown that the riboflavin content of the livers of embryos is decreased after treatment with boric acid.

Spleen.—The size of the spleen is diminished, or its growth inhibited by X-rays and the polyfunctional alkylating agents (nitrogen mustard and related compounds). It is sometimes enlarged in cobalt-treated embryos. The size and appearance of the spleen may indicate alterations in hematopoietic activity.

Blood.—Pale embryos have been produced by 2,6-diaminopurine, cobalt, azaserine, X-rays, urethane, and boric acid. Paleness suggests anemia, although this has been demonstrated only with 2,6-diaminopurine. Routine studies on the peripheral blood of treated chick embryos and on the levels of hematopoietic activity in the liver, spleen, and bone marrow would be of great interest.

Edema.—Several forms of edema are found in the chick embryo. Edema may be generalized, and the fluid is clear, milky, or gelatinous and may be localized in large unilateral or bilateral rump sacs. Edema has occurred frequently after physostigmine, ethionine, selenium, cortisone, deoxyguanosine, and X-rays. Important work remains to be done on the mechanism of edema formation, and its relation to renal, hepatic, cardiac, electrolyte, or hematologic disturbances.

Feather growth.—Feather tracts and feather growth have been carefully studied in the chicken. In the chick embryo, some agents cause bald areas, interference with feather growth, stunting of feathers, and clubbed down and thin, pale feathers. These changes have been seen after administration of cortisone, boric acid, azaserine, sulfanilamide, deoxyguanosine, and ethionine. Landauer (18) has observed restriction in melanin formation in Black Minorca embryos treated with insulin, boric acid, or pilocarpine.

We have not examined other systems, such as the heart, urinary, and intestinal tracts, gonads, muscles, and endocrine organs, but if these were studied appropriately, important alterations would undoubtedly be found.

The use of the chick embryo to detect drugs with potential teratogenic activity in man is an important point for discussion. The chick embryo has responded with specific developmental disturbances to a number of growth-inhibiting drugs with known growth-inhibiting activity and teratogenic effects in other species (38). On the other hand, thallium and deoxyguanosine are potent teratogens in the chick embryo, but similar effects have not been observed in mammals. Presumably in mammals, the mother may detoxify the drug before it reaches the embryo in sufficient quantity, or it may not penetrate the placenta. Thalidomide, in our experience, has had no significant teratogenic activity in the chick embryo, even when we have attempted to sensitize the embryo to a micromelia-producing drug by the injection of insulin. Yet its teratogenic effect in rabbits and man is well established.

Thus, while the chick embryo is an interesting system in which to study drug effects, undue reliance should not be placed on it as a major predictive system for the teratogenic activity of a drug in man. The chick embryo is useful for studying the growth-inhibiting action of certain drugs, and in studying their mechanism of action. The results obtained in this system, for the moment, are relevant only to the avian embryo, and there must be many reservations in extrapolating them to other species.

SUMMARY AND CONCLUSIONS

1. The chick embryo is a useful system for examining the effects of drugs on growth and development. It has been shown that specific drugs can produce specific disturbances in development which are detectable by gross

examination. These effects are consistent and reproducible from experiment to experiment.

2. The embryo appears to be most responsive in the majority of drugs studied when injected on the fourth day of incubation.

3. Major adverse effects of drugs on the developing embryo include stunting in growth, anemia, generalized edema, feather inhibition, corneal cysts, defective eyelids, fatty liver, atrophy of spleen, brain hemorrhage and necrosis, inhibition of muscle growth, and a variety of skeletal deformities including facial coloboma and cleft palate, fused vertebrae, fusion of bones of the extremities, and absence of bones of the extremities, duplications of bones, and inhibition of osteogenesis. These effects can be detected by gross examination and in selected cases by clearing the embryo so that a detailed examination of the skeleton is possible.

4. There are several characteristic syndromes that have been related to certain classes of drugs. The ones described in this report include (a) micromelia and parrot beak, (b) fusion of vertebrae (wry neck), (c) stunting of growth with facial coloboma and cleft palate, and inhibition of skeletal development with bony deletions and fusions, (d) stunting of growth with feather inhibition and disturbances in osteogenesis, (e) fatty liver and inhibition of muscle development, and (f) hemorrhage and necrosis of brain.

5. The chick embryo cannot reliably predict which drugs will be teratogenic in mammals, including man. Certain growth-inhibiting drugs have been consistently effective in the chick and mammalian systems, others have not. Thalidomide, for example, has had no consistent effect on the chick embryo when given at maximum tolerated doses from 0 to 8 days of incubation.

REFERENCES

1. ROMANOFF, A. L. AND ROMANOFF, A. J. *The avian egg*. New York: John Wiley & Sons, Inc., 1949.
2. HAMILTON, H. L. (ed.) *Lillie's development of the chick: An introduction to embryology*. 3d ed. New York: Henry Holt and Co., 1952.
3. NEEDHAM, J. *Biochemistry and morphogenesis*. Cambridge: Cambridge University Press, 1942.
4. WILLIER, B. H., WEISS, P. A., AND HAMBURGER, V. (eds.) *Analysis of development*. Philadelphia: W. B. Saunders Co., 1955.
5. MINER, R. W. (ed.) The chick embryo in biological research. *Ann. N.Y. Acad. Sc.* **55:** 37, 1952.
6. LANDAUER, W. The hatchability of chicken eggs as influenced by environment and heredity. Monograph 1., Univ. of Connecticut, Agricultural Experimental Station, Storrs, Conn., 1961.
7. ROMANOFF, A. L. *The avian embryo: Structural and functional development*. New York: The Macmillan Co., 1960.
8. LANDAUER, W. On the chemical production of developmental abnormalities and of phenocopies in chicken embryos. *J. Cell. Comp. Physiol.* **43**(Suppl.) : 261, 1954.

9. Stock, C. C. Aspects of approaches in experimental cancer chemotherapy. *Am. J. Med.* **8**: 658, 1950.

10. Karnofsky, D. A. Investigation of diverse systems for cancer chemotherapy screening. XVI. Assay of chemotherapeutic agents on the developing chick embryo. *Cancer Research* **15**(suppl. 3): 83, 1955.

11. Landauer, W. Genetic and environmental factors in the teratogenic effects of boric acid on chicken embryos. *Genetics* **38**: 216, 1953.

12. Franke, K. W., Moxon, A. L., Poley, W. E., and Tully, W. C. Monstrosities produced by the injection of selenium salts into hens' eggs. *Anat. Rec.* **65**: 15, 1936.

13. Karnofsky, D. A., Ridgway, L. P., and Patterson, P. A. Growth-inhibiting effect of cortisone acetate on the chick embryo. *Endocrinology* **48**: 596, 1951.

14. Ridgway, L. P. and Karnofsky, D. A. The effects of metals on the chick embryo: Toxicity and production of abnormalities in development. In: *The chick embryo in biological research. N.Y. Acad. Sc.* **55**: 203, 1952.

15. Karnofsky, D. A., Stock, C. C., Ridgway, L. P., and Patterson, P. A. The toxicity of vitamin B_6, 4-desoxypyridoxine, and 4-methoxy-methyl pyridoxine, alone and in combination to the chick embryo. *J. Biol. Chem.* **182**: 471, 1950.

16. Fugo, N. W. Effects of hypophysectomy in the chick embryo. *J. Exper. Zool.* **85**: 271, 1940.

17. Karnofsky, D. A. and Ridgway, L. P. The production of injury to the central nervous system of the chick embryo by lead salts. *J. Pharmacol. Exper. Therap.* **14**: 176, 1952.

18. Landauer, W. Abnormality of down pigmentation associated with experimentally produced skeletal defects of chicks. *Proc. Nat. Acad. Sc.* **39**: 54, 1953.

19. Dagg, C. P. and Karnofsky, D. A. Some effects of 6-aminonicotinamide in chick embryos. *Fed. Proc.* **17**: 361, 1958.

20. Landauer, W. Niacin antagonists and chick development. *J. Exp. Zool.* **136**: 509, 1957.

21. Murphy, M. L., Dagg, C. P., and Karnofsky, D. A. Comparison of teratogenic chemicals in the rat and chick embryos and exhibit with additions for publication. *Pediatrics* **19**: 701, 1957.

22. Landauer, W. On teratogenic effects of pilocarpine in chick development. *J. Exper. Zool.* **122**: 469, 1953.

23. Landauer, W. Insulin-induced abnormalities of beak, extremities, and eyes in chickens. *J. Exper. Zool.* **105**: 145, 1947.

24. Zwilling, E. and DeBell, J. T. Micromelia and growth retardation as independent effects of sulfanilamide in chick embryos. *J. Exper. Zool.* **115**: 59, 1950.

25. Dagg, C. P., Karnofsky, D. A., Stock, C. C., Lacon, C. R., and Roddy, J. Effects of certain triazenes on chick embryos and on tumors explanted to the chorioallantois. *Proc. Soc. Exp. Biol. Med.* **90**: 489, 1955.

26. Karnofsky, D. A. The use of the developing chick embryo in pharmacologic research. *Stanford Med. Bull.* **13**: 247, 1955.

27. Karnofsky, D. A., Ridgway, L. P., and Patterson, P. A. The production of achondroplasia in the chick embryo with thallium. *Proc. Soc. Exper. Biol. Med.* **73**: 255, 1950.

28. Karnofsky, D. A., Ross, C., and Leavitt, C. G. W. Effects of quaternary ammonium and cholinergic drugs on skeletal development of chick embryo. *Fed. Proc.* **13**: 373, 1954.

29. Karnofsky, D. A. and Clarkson, B. Cellular effects of anticancer drugs. *Ann. Rev. Pharmacol.* **3**: 357, 1963.

30. Dagg, C. P. and Karnofsky, D. A. Teratogenic effects of azaserine on the chick embryo. *J. Exper. Zool.* **130**: 555, 1955.

31. Dagg, C. P., Karnofsky, D. A., Lacon, C., and Roddy, J. Comparative effects of 6-diazo-5-oxo-L-norleucine and O-diazo-acetyl-L-serine on the chick embryo. *Proc. Amer. Assoc. Cancer. Res.* **2**: 101, 1956 (Abst.).

32. Karnofsky, D. A. and Lacon, C. Effects of physiological purines on the development of the chick embryo. *Biochem. Pharmacol.* **7**:154, 1961.

33. KARNOFSKY, D. A. AND LACON, C. R. Protection against deoxyguanosine (dGUR) toxicity in the chick embryo with deoxycytidine (dCR). *Fed. Proc.* **21**: 379, 1962 (abstr.).
34. KARNOFSKY, D. A., MURPHY, M. L., AND LACON, C. R. Comparative toxicologic and teratogenic effects of 5-fluoro-substituted pyrimidines in the chick embryo and pregnant rat. *Proc. Amer. Assoc. Cancer Res.* **2**: 312, 1958 (abstr.).
35. MOSCONA, M. H. AND KARNOFSKY, D. A. Cortisone-induced modifications in the development of the chick embryo. *Endocrinology* **66**: 533, 1960.
36. KARNOFSKY, D. A., DAGG, C. P., AND LACON, C. R. Effects of ethionine and related compounds on the developing chick embryo. *Fed. Proc.* **14**: 438, 1955.
37. LANDAUER, W. Complex formation and chemical specificity of boric acid in production of chick embryo malformations. *Proc. Soc. Exper. Biol. Med.* **82**: 633, 1953.
38. KARNOFSKY, D. A. Influences of antimetabolites inhibiting nucleic acid metabolism on embryonic development. *Trans. Ass. Amer. Physicians* **73**: 334, 1960.

CHAPTER 9

EFFECT OF PROTEINS, ANTIBODIES, AND AUTOIMMUNE PHENOMENA UPON CONCEPTION AND EMBRYOGENESIS

ROBERT L. BRENT

The theoretical classification of human congenital malformations can be divided into four arbitrary categories.

1. Environmental or postconception factors (drugs, irradiation, maternal disease, maternal infection, some chromosome abnormalities, etc.).

2. Genetic and/or preconception factors (hereditary anomalies, somatic germ cell mutations, some chromosome abnormalities).

3. Combinations of 1 and 2.

4. Errors of growth and differentiation in the embryo which are unrelated to exogenous or hereditary genetic factors, but due only to the statistical probability that a small percentage of embryos will fail to accomplish the intricate processes that are necessary for normal differentiation. Thus, it is possible that malformations may result from predictable but rare errors in the biology of embryonic growth and differentiation, similar to the concept of a predictable mutation rate at a gene locus. If this is the case, and if the majority of malformations fall into this category, then the prevention of congenital malformations will be a most difficult task, albeit not impossible.

Although environmental factors may account for a small portion of the total incidence of human malformations, it is the category in which the most successful preventive measures can be instituted. It is for this reason that so much attention has been devoted to the study of exogenous factors or the altered maternal environment of the fetus.

It has been known for decades that every pathological alteration in the physiology of the maternal organism does not result in sterility, or if an embryo was present, teratogenesis. Thus, investigators have attempted to understand why one altered metabolic state would affect reproductive function while another would not; why two closely related drugs were not both teratogenic. One new area deals with the relationship between the occur-

Supported by NIH Grant HD-00630.

Professor of Pediatrics and Radiology, and Director, Eleanor Roosevelt Research Institute, Jefferson Medical College, Philadelphia, Pennsylvania.

rence of immunologic disease and altered states of protein metabolism in the maternal organism and the following pathology:

1. Some aspects of infertility and abortion.

2. Postpartum pathology resulting from the presence of immunopathological states during pregnancy.

3. Abnormal embryonic development.

It is the purpose of this paper to review the important areas of research in this field and to discuss the validity and significance of some of the investigations and resulting hypotheses. Furthermore, whenever applicable, the actual or potential clinical significance of these findings will be stressed.

CLINICAL AND EXPERIMENTAL SIGNIFICANCE OF ANTIBODIES AGAINST GAMETOCYTES, GONADS, AND HORMONES

Attempts to induce sterility by immunization techniques were reported 30 to 40 years ago (1–3). Aspermatogenesis can be induced in the male animal by the injection of either heterologous or homologous testicular tissue or sperm (4–7). In fact one investigator produced seminiferous tubule degeneration in one testis by injecting Freund's adjuvant into the foot pad of a guinea pig and turpentine into the other testis (8). The onset of pathological changes in the progenitive tissue is quite rapid, being reported as early as 5 days after the first injection of Freund's adjuvant and the testicular tissue. The experimental production of sterility in mammals from an autoimmune or immune response to male progenitive tissues has been described histologically and proven immunologically, as evidenced by the fact that immunization with sperm antigen will fail to produce aspermatogenesis if the animal is immunologically depleted by removal of the thymus (9). It is also interesting that demonstration of sperm antibodies in the serum of men is not necessarily associated with aspermatogenesis or sterility (10).

Sterility can be induced in the female *rodent* by repeated immunization with male testis or sperm (2, 11–14). Sperm antibodies in the female are fixed in the guinea pig uterus (15, 16) and can be demonstrated in the serum and in the secretions of the genital tract when the female animal is immunized parenterally. The sperm antibodies are present in the genital tract first when the immunizing exposure is only via this route (17, 18). Circulating antibodies to sperm or testicular tissue can be produced via transvaginal instillation if Freund's adjuvant is combined with the antigen (19). It is also interesting that the pregnant rabbit is more readily immunized wth bull spermatozoa via the genital tract than the non-pregnant rabbit (18). Experimental sterility produced in the female by immunologic techniques may result from the interference of sperm migration, the prevention of fertilization or implantation, or the rejection of the implanted embryo.

Sterility and infertility is a multifaceted problem in the human, and the relative importance of immunopathological phenomena is as yet unknown, although it has been studied for many years. Antibodies to the sperm antigens have been found in the sera of men and women with an infertility problem (20, 21), but so too have they been demonstrated in fertile couples (10, 13, 14). There appears to be a higher incidence of these antibodies in infertile couples than in the fertile couples, but it is not known whether this finding is significant.

Considering the possibility that male tissues may sensitize the female during pregnancy, Renkonen (22, 23) and his associates postulated that a successful male pregnancy is less likely to occur in women who have previously delivered a male offspring, since male pregnancies may immunize a small portion of mothers against male Y chromosome antigen and this immunization may be harmful to subsequent male fetuses. Renkonen reports a progressive change in the sex ratio with parity that supported his theory (23). Edwards (24) re-evaluated Renkonen's data and did not find any significant change in the sex ratio with increasing parity. McLaren (25) attempted to test this theory in mice and found a sex ratio of 102 males/106 females in offspring of immunized mice. She stated, "It therefore seems unlikely, though of course not impossible, that immunization against Y chromosome antigen plays the part postulated for it by Renkonen."

Sutherland and Landing (26) studied an experiment of civilization rather than the usually described "experiment of nature" when they investigated the etiology of infertility among prostitutes. They found no evidence of antibodies against sperm or other testicular elements when the sera of prostitutes were tested with fluorescent labeling techniques.

Antisera have been prepared against some of the pituitary and placental hormones which are important in controlling the estrus cycle and maintaining pregnancy (27). For example, a new pregnancy test utilizes an antiserum to chorionic gonadotropin (28, 29). Immunological studies of hormones may also determine whether the active hormones secreted by tumors are different from hormones secreted by the normal glands. As an example, it has been demonstrated that chorionic gonadotropin secreted by the normal placenta and a choriocarcinoma are identical (30). There have also been several attempts to produce pathological states utilizing these antibodies. Antiserum to interstitial cell-stimulating hormone has been reported to suppress ovulation and affect the estrus cycle. The effects of antisera made against follicle-stimulating hormone and chorionic gonadotropin have also been investigated (31). The results of these investigations on the effects and uses of antihormones may yield valuable information regarding the specific functions of these hormones and their sites of production in the pituitary. Fluorescein or isotope labeling of these antisera may reveal the sites of localization of these hormones in the peripheral tissues. The knowledge obtained from observing the suppressive effects of some of these antisera and the peripheral sites of localization of these antisera may add to our understanding of sterility and provide new approaches to simplifying the problem of contraception in humans.

Whether the subject of sperm, testicular, ovum, and reproductive hormone antibodies turns out to be important in solving some of the pathologic states of human reproduction is as yet undetermined, but they have yielded some interesting experimental models for studying infertility and some very provocative theories.

Immunopathologic Significance of the Transfer of Cells, Antigens, Antibodies, and Proteins between Mother and Fetus

The relationship between the mammalian mother and fetus affords both organisms the temporal opportunity to share each other's products of anabolism and catabolism and, in some circumstances, even the viable and non-viable products of replicating tissues. Since we are concerned with the immunologic aspects of the fetal-maternal relationship, the entire subject of placental transfer will not be considered. The following discussion will be limited to the passage of maternal and fetal *cells* and *proteins* through the placenta.

Newborn infants are susceptible to certain infections regardless of their mothers' immune status, and these same infants are resistant to other infections, providing their mothers have been previously immunized. The correct inference, that all immune bodies do not cross the placenta, was therefore derived from clinical data. The present status of this field can be summarized as follows:

1. The immunity to many infections can be passively transferred to the fetus (32–37).

2. High, naturally occurring titers of antibody in the mother, or titers increased by artificial active immunization, are reflected in the serum antibody titers in the fetus (33, 38).

3. The transfer of serum proteins between mother and fetus is a two-way affair, although equilibrium may not be reached (37, 39).

4. The qualitative nature of the protein interchange and the quantitative rate of interchange is constantly changing throughout the gestation. Furthermore, the route of transfer, i.e., placental versus extraembryonic membranes and cavities, varies in different species groups (34, 37, 40). Since the fetal capacity to produce proteins is constantly improving, it is very difficult to analyze the dynamics of the protein interchange between mother and fetus.

5. An analysis of the fetal serum proteins reveals that the spectrum of immune proteins in the mother is not exactly mirrored in the fetus. Gitlin, Rosen, and Michael (41) have shown that the 7 S gamma$_2$ globulins are present and that there is a deficiency of the 19 S gamma$_1$ globulins in the newborn infant. The antibodies to many enteric bacteria are believed to be present in the 19 S gamma$_1$ globulin fraction. These investigators attribute the newborn infant's susceptibility to gram-negative enteric coli infections to the lack of passively transferred 19 S gamma$_1$ globulins. Other discrepancies between the immune proteins in the mother and fetus may be immunologically significant in areas that we are unaware of at the present time. For instance, Freda and Carter (42) have demonstrated a difference in the permeability of beta isoantibody subgroups A$_2$ and A$_1$, in that A$_2$ is more readily transferred by the placenta when the mother is group O or A$_2$ than when she is A$_1$. This indicates that variations of protein transfer via the placenta may be partially determined by genetic constitution and that individuals may differ considerably in their ability to transfer immune bodies and protein to the fetus.

Protective antibodies are not the only antibodies which traverse the placenta, and many immunopathologic proteins or toxic proteins which produce disease in the mother can be passively transferred to the fetus. The following cases are examples of this situation:

1. The L.E. phenomenon of lupus erythematosus can be passively transferred from mother to infant (43–45). In some instances dermatologic lesions similar to the mother's have appeared in the newborn infant only to disappear during the neonatal period (46).

2. The symptoms of thyrotoxicosis and myasthenia gravis can be present in newborn infants of mothers manifesting signs of these illnesses.

3. Hemolytic anemia in the fetus (erythroblastosis) is produced by maternal sensitization to fetal red blood cells.

4. Thrombocytopenia may be present in an infant born to a mother with acute or chronic thrombocytopenia or *may* result from maternal sensitization to fetal thrombocytes (47).

5. Agranulocytosis may occur following maternal sensitization to fetal leukocytes (48–50).

6. Autoimmune disease in the mother which might result in fetal loss or malformation will be discussed in the last section, but is also listed here since this immune response may affect the fetus.

From the foregoing it is obvious that at least three fetal cellular materials reach the maternal circulation: red blood cells, white blood cells, and platelets. The incidence of sensitization to fetal white blood cells and platelets resulting in clinical disease is low, but the incidence and quantity of exposure are unknown. On the other hand, a great deal has been written about red blood cell migration into the maternal circulation (51–53). In fact Brown and Cowels (54) estimate that 44 per cent of the pregnant women they studied had significant numbers of fetal erythrocytes in their circulation.

The appearance of trophoblastic tissue in the maternal circulation and organs is another interesting facet of the subject of fetal tissue migration to the mother (55). It has been estimated that at least one trophoblast cell is present in each milliliter of venous blood leaving the uterine circulation (55). Investigators have demonstrated placental antigens in the maternal circulation (56, 57). The presence of placental antigens and trophoblastic tissue in the maternal circulation has not gone uninterpreted. Thomas, Douglas, and Carr (56) postulated "that the invasion of maternal blood by the trophoblast plays a comparable role in maintaining a state of desensitization or immunologic unresponsiveness by exposing the mother to a constant excess of fetal antigen." This theory does not agree with the experimental results of Simmons and Russell (58), which indicated that the trophoblast was an immunologic barrier between the fetus and maternal organism. On the other hand, it is difficult for any observer or investigator to attribute no purpose or meaning to trophoblastemia or trophoblast antigenemia. It is possible that trophoblastemia may be the primary phenomenon responsible for Y antigen immunization as postulated by Renkonen's theory (22, 23) which was alluded to in the first section. It is also possible that the quantity of trophoblastic tissue in the maternal circulation may be responsible for the initiation of parturition. Substances in the trophoblast may have hormonal functions whose absence in the postpartum period initiates some of the normal involutional processes. It is obvious from the foregoing that the true importance of trophoblastemia is not known.

There is no doubt that cellular material migrates from the fetus to the mother; the reverse phenomenon may be more important but is more difficult to document. In some instances of malignancy in a pregnant mother, the tumor has appeared in the newborn infant during the neonatal period or has, on occasion, been present at birth (59, 60). The malignant melanoma seems to have a propensity for metastasizing to the placenta or fetus, while other malignancies of the mother rarely involve the fetus. The transfer of malignant cells, or the agent which induces malignancy, from mother to fetus is readily proved by the finding of the maternal tumor in the fetus, combined with the rarity of the occurrence of spontaneous neoplasms in the newborn infant.

On the other hand, the presence of normal maternal cells in the fetus is not as easily documented, and yet it would be of great interest to know whether red blood cells, white blood cells, platelets, or their precursors leave the maternal circulation and reach the fetus. This may be extremely important, since surviving adult cells in the fetus could result in (1) tolerance to certain foreign antigens, (2) sensitivity to these same antigens, or (3) possibly a graft-versus-host reaction, if enough immunologically competent adult cells survive in the fetus. As an example, Halasz and Orloff (61) sensitized pregnant rabbits with a graft from an unrelated doe. Sensitization to the graft was reported in the fetus, and these authors attributed these results to the passage of circulating antibody through the placenta, followed by reattachment to fetal lymphocytes. If their observations are correct, sensitization might also be explained by the *passage of maternal lymphocytes* to the fetal circulation. Not only might the phenomenon of the transfer of competent adult cells to the embryo explain some of the syndromes of growth failure and unexplained death in infancy (runt disease), but it might also offer an explanation for the *etiology* of autoimmune disease, the *familial incidence* of autoimmune disease, and the *variable onset* of these diseases. This concept certainly warrants further investigation, since the ability to acquire adult cells *in utero* might depend on maternal constitution and the severity and onset of the symptoms might be related to the dose of competent cells reaching the fetus.

EFFECT OF ANTIBODIES AND PROTEINS
UPON EMBRYONIC DEVELOPMENT

Reports as early as 1918 deal with the production of malformations with tissue antiserum. The early investigations of Guyer and Smith (62, 63) proved be in error but were the basis of many more investigations. Guyer and Smith reported that not only did lens antiserum produce eye malformations when injected into pregnant rabbits but that these eye malformations were hereditary. Attempts to confirm the teratogenic and mutagenic aspects of lens antiserum were unsuccessful (64–68) until Miller (69) reported that he obtained three eye malformations in 459 live offspring following the injection of lens antiserum into pregnant rabbits. Miller could not substantiate the hereditary claims made by Guyer and Smith. The three malformations that Miller reported were a peculiarly shaped iris, microphthalmia, and cloudiness of the eye. There was one malformation in 284 control offspring. It appears that *neither* of Guyer and Smith's original findings were corroborated by Miller's study. On the other hand, almost all the studies did indicate an increase in embryonic mortality following passive immunization with the lens antiserum. Thus, although this specific antiserum may not have teratogenic properties when injected into the pregnant doe, it may have deleterious effects on the viability of the embryo. Although this line of investigation began over 50 years ago, most of the research has been undertaken in the last two decades and has been concerned with mammalian embryology and chick and amphibian embryology in about equal proportion.

Studies with non-mammalian organisms.—Although the immunologic studies dealing with mammalian embryos may more likely mimic situations that can occur in the human, the chick and amphibian embryos provide experimental conditions that are unavailable in the mammal. As an example, the embryologist has had a great deal more success in utilizing lens antiserum in the chick than in the mammal, since the antiserum can be injected directly into or upon specific areas of the embryo and the concentrations attained can be readily controlled. Lens antiserum has demonstrated its effectiveness in interfering with lens development, eye development, and, if the concentration was high enough, with the viability of the entire embryo (70–73). Langman (72) demonstrated that the antibody responsible for lens degeneration was not against the beta or gamma crystallin but against the alpha crystallin of the lens. Furthermore, not only was this antigen-antibody reaction limited to certain ectodermal tissues but its cytopathogenic effect was limited to certain stages of gestation, since the chick embryo was resistant to the effective dose of antiserum after the 16-somite stage (73).

Ebert evaluated the effect of tissue antisera on the development of the predifferentiated embryo (74). Johnson and Leone (75) used chick serum antiserum and actomyosin antiserum to inhibit growth in the chick embryo. Clayton (76) used tissue antisera which were adsorbed with tissue antigens or serum factors to produce abnormalities in the newt. The unabsorbed antiserum usually resulted in the death of the embryo. Nace and Inoue (77) reported cytolysis following exposure of embryos to antineurula serum. Thus, alterations in growth and differentiation have been reported in species where antibodies can be applied directly to the embryo.

Although the effects of specific tissue antisera upon the developing embryo may have been predictable, this was not the case with tissue antigens. Certain fractions of adult chicken brain interfere with the development of the chick embryo brain when the embryo is injected with small quantities of brain antigen (78–80). Furthermore, brain antigen also interferes with normal eye development. The exposure of the embryo to similar quantities of chicken muscle, heart, kidney, skin, or mammalian brain does not cause specific malformations of the chick brain (78). Langman contends that a soluble antigen which is present only in brain tissue is responsible for the teratogenic effect (80). Therefore, whatever the mechanism of teratogenesis of both tissue antisera or tissue antigens, there does appear to be some selectivity and specificity to their toxic properties.

Other investigators have exposed the chick embryo to the sera of patients with various diseases in order to uncover altered metabolic states in the human patients. Sacerdote de Lustig and Fiszer (81, 82) studied the effect of serum obtained from cancer patients on the development of the chick embryo. The serum was injected intravenously on the ninth or eleventh day, and the embryos were examined 4 to 6 days later. The whole serum, fractions IV, V, VI, and gamma globulin were obtained from patients suffering with various malignant tumors or from patients following various forms of cancer therapy. The incidence of malformations was less than 2 per cent in chick embryos receiving normal sera, sera from patients with benign tumors, sera from patients treated for their cancer, and fractions IV, V, and VI from all sources. Whole sera from cancer patients produced a 52 per cent incidence of malformations and the gamma globulin fraction of this serum produced a 65 per cent incidence of malformations.

The authors postulate that an enzyme or virus may be the etiologic factor since incubation at 60° C. destroys the teratogenic properties of the serum. If it were an antibody, it should resist treatment at 60° C. Furthermore, a virus would be unlikely to be present only in the gamma globulin fraction. The most difficult portion of this work to understand is that the serum from treated cancer patients no longer is teratogenic. The last statement the authors made in their paper is that they are attempting to prevent the malformations by using "anticancer" serum in conjunction with the teratogenic serum (82). Other investigators have also attempted to use human cancer sera as a teratogenic agent. Cholewa (83) and Lacon and Karnofsky (84) found that cancer sera were ineffective as a teratogenic agent in chick embryos and that the control and experimental chick embryo had the same incidence of malformations. Although some of the basic embryologic information gathered from these studies can be applied to the mammal, the teratogenic findings cannot be readily applied since the administration of antiserum to the pregnant mammal results in an experimental situation considerably different from the application of antiserum to the chick or amphibian embryo.

Studies with the mammalian organism.—This general interest in the *effect of proteins on embryonic development* is reflected in a number of interesting but diverse papers dealing with the evaluation of protein metabolism during pregnancy or the effect of administered proteins upon embryonic development. Langman, Van Drunen, and Bowman (85) performed electrophoretic analyses on the serum of pregnant women and concluded that certain electrophoretic patterns could be of prognostic value in determining the normalcy of the fetus. Gillman and co-workers (86) reported the production of congenital malformations with trypan blue in 1948. They postulated a number of mechanisms to explain the teratogenic action of trypan blue, one of which was an induced alteration in protein metabolism (87, 88). Ochiac (89) reported that the globulin fraction of the human placenta, when injected into pregnant rabbits, resulted in the histologic changes resembling toxemia of pregnancy. Law and Law (90) warned against the use of prophylactic gamma globulin in pregnant women because they observed a high incidence of abortions in pregnant women treated with gamma globulin. Thus, there are a number of papers alluding to the toxicity of heterologous or isologous serum proteins to the embryo or embryonic site. These findings are difficult to interpret since most of the investigations are quite recent and the characterization of the teratogenic or abortogenic factors has not been completed. Further investigation of the maternal and fetal protein metabolism in normal and abnormal pregnancies is certainly indicated.

In 1940 Seegal, Loeb, and co-workers began an interesting series of experiments in which rabbit antirat-placental serum was administered to pregnant rats (91–94). In later years they switched species and utilized dog placenta as the antigen (95, 96). Other investigators have reproduced Seegal's experiments utilizing various species (97, 98). The sum of these findings indicates that both antikidney and antiplacental sera induce an identical form of chronic nephritis. When this antiserum was injected into pregnant rats, hypertension, nephritis, albuminuria, and edema were produced, mimicking the clinical syndrome of "toxemia of pregnancy." Fetal loss was increased as evidenced by resorptions or stillbirths, depending on the time in gestation that antiserum was injected. No malformations were reported in these experiments in spite of the fact that the antiserum was administered as early as the ninth day of gestation. Pressman and Korngold (99) evaluated the immunologic relationship between rabbit antirat kidney antiserum and Seegal's antiplacental serum. They concluded that, although many of the localizing antibodies of kidney and placental antiserum are in common, different antibodies are concerned with the interruption of pregnancy and the production of nephrotoxicity. Using fluorescent antibody techniques, Steblay (100–102) demonstrated that antiplacental serum localizes in the glomerular basement membrane, tubules, capsules, capillaries, media, and adventitia of the arteries in the kidney. He also produced an autoimmune nephritis in pregnant sheep which did not affect the kidneys of the fetus during pregnancy or during the nursing period. This autoimmune nephritis was severe enough to be fatal to the mothers in from 45 to 54 days from the onset (102).

Gluecksohn-Waelsch (103) reported that active autoimmunization of female mice with brain antigen resulted in central nervous system malformations in approximately 10 per cent of the offspring. Furthermore, similar immunization with heart antigen did not result in malformations of the central nervous system. She concludes, "it appears likely that the abnormalities of the nervous system observed in the present experiments are due to a direct effect of maternal brain antibodies on the developing nervous system of the embryo." In discussing her experimental results, she indicates that she has not eliminated brain antigen as the potential teratogen and that some of the "malformed" embryos came from litters where no antibody was demonstrated in the mother. Other criticisms of this work are that the embryos were studied very early in gestation (tenth day) at which time the central nervous system is the major portion of the embryo and that any dying embryo is likely to have distortions or retardation of its major structures. It would have been exceedingly helpful if some of these litters were allowed to go to term. Second, there was no evidence that any antibody reached the embryo. Third, no encephalitic symptoms were reported in the maternal organisms. Finally, it was reported that the embryonic mortality from maternal immunization with heart tissue was 10.7 per cent (23/185) and from maternal immunization with brain tissue was 10.9 per cent (28/256). Since agents which produce malformations frequently increase fetal mortality, it is surprising that the control and experimental mortalities were the same. On the other hand, this provocative study has been repeated, and although no brain malformations were seen in the offspring, eye malformations were observed (104). The main concern with Glucksohn-Waelsch's work is not whether malformations can or cannot be made by antibodies or immunizing procedures, but whether organ specific antibodies or immunizations in the mother can result in specific organ malformation in the embryo.

In the course of the elegant experiments of Levi-Montalcini, dealing with the effects of a nervous tissue growth-stimulating protein, she prepared antiserum against this protein (105). This antiserum greatly reduced the normal growth of sympathetic nervous tissue in neonatal and adult animals, but it had no effect upon the embryo or fetus when the antiserum was administered to the pregnant maternal organism (105). Apparently, the factor responsible for interfering with the nerve growth-stimulating protein did not cross the placenta or was utilized entirely by the mother.

In 1961, Brent, Averich, and Drapiewski (106) reported that the administration of rabbit antirat kidney antiserum to pregnant rats resulted in congenital malformations. Similar results were obtained in 1963 by David, Mercier-Parot, and Tuchmann-Duplessis (107). The types of malformations produced by this antiserum were similar to the effects of other teratogenic agents, such as X-irradiation, trypan blue, or uterine vascular clamping (106, 108, 109). The administration of this antiserum resulted in a different spectrum of malformations on various days of gestation. Although some malformations were produced when the teratogenic antiserum was administered to pregnant rats during several stages of gestation, other malformations were produced on only a particular day of gestation (107). The incidence of malformations was dose dependent and the dose of antiserum could be adjusted to produce malformations in 100 per cent of the offspring. A further increase in the dose would result in 100 per cent resorptions. The kidney antiserum demonstrated all the properties of a general teratogenic agent, namely, (1) growth retardation, (2) malformation, and (3) embryonic lethality (106, 109). Although these three effects were produced by the usual working dose of kidney antiserum, it was possible to produce malformed offspring with low doses of antiserum that neither caused significant fetal death nor fetal growth retardation (109). This finding, along with the knowledge that the teratogenic agent is a protein, distinguishes kidney antiserum from other laboratory teratogenic agents. This antiserum is more potent as a teratogenic agent than are many of the cancer chemotherapeutic agents. Second, this is a potent experimental teratogen which is the metabolic product of a mammal rather than a synthetic agent produced by an organic chemist or an alkaloid derived from some rare plant. The qualifications or uniqueness of this teratogenic agent in no way indicates its mechanism of action which includes (1) a metabolic lesion in the maternal organism, (2) direct action upon the embryo, or (3) interference with the decidual reaction or the trophoblast (110). The fact that placental antiserum has also been found to be teratogenic, that kidney antiserum interferes with the growth of rat placenta when grown in tissue culture (111), and that kidney antiserum localizes in the kidney, adrenal, and placenta, may indicate that kidney antiserum in some way interferes with the function of the deciduotrophoblast; but this is only conjectural. Further investigations are necessary to explain the teratogenic mechanism of kidney and placental antiserum.

The present clinical approaches to immunologically induced teratogenesis are related more to the laboratory work of Gluecksohn-Waelsch (103) and Barber, Willis, and Afeman (104) than to the work of Brent (106, 109, 110), since the former have been concerned with the production of autoimmune disease in the mother and the induction of specific organ malformations in the fetus.

The association between autoimmune disease and congenital malformations was brought into focus by the investigations into the etiology of cretinism. An increased incidence of thyroid antibodies in the mothers of infants with cretinism has been described, although mothers with thyroid antibodies in their sera did not necessarily have infants with cretinism (112). These antibodies cross the placental barrier and are present in normal infants and infants with cretinism (112, 113). Chandler, Kyle, Hung, and Blizzard (114) attempted to mimic this clinical situation in experimental animals and, although they were able to demonstrate thyroid antibodies in the fetus, no thyroid disease occurred in the fetus. These results are difficult to explain if one believes that the thyroid antibodies in the serum are responsible for the thyroid pathology.

Hellwig and Wilkinson (115) questioned the actual concept of autoimmune thyroiditis and reported, "Our experiments suggest that chronic thyroiditis is a defense reaction against iodinated lipids and that immune bodies are not the cause but the effect of thyroiditis." Recently, hereditary Addison's disease has been placed in this same category as some forms of thyroiditis (116).

In clinical instances of autoimmune disease and experimentally induced autoimmune diseases, the pathological states in various organs and tissues are predominantly dependent on the infiltration of sensitized lymphocytes in spite of the fact that circulating antibodies may be present. Therefore, since it is unlikely that the sensitized lymphocytes can cross the placenta in large numbers, one would not expect the maternal autoimmune disease to be consistently reflected in the fetus unless the circulating antibody has some pathologic effect of its own. Thus, the experimental results (103, 104, 114) and the theoretical application of our understanding of autoimmune disease would lead us to discount an autoimmune process as a good experimental method of producing specific malformations or an important factor in the etiology of abnormal human offspring. This does *not* mean that circulating antibodies, such as teratogenic kidney antibody, *are* related to the problem of human malformations; however, both these methods of protein-induced malformations need to be more extensively investigated. For as long as there exists the present uncertainty as to the etiology of most malformations, no one can predict which avenue of research will yield results that may aid in the prevention of human malformations.

SUMMARY

1. The experimental immunization of male and female animals with sperm or testicular material can unequivocally produce infertility and sterility in isologous or heterologous animals. As a result of this line of experimental work, a number of provocative concepts and theories have been promulgated, including attempts to explain sterility in prostitutes and Renkonen's theory of maternal Y antigen sensitization as a cause of the decreasing sex ratio with parity. At the present time, even though sperm and testicular antibodies have been found in men and women, their relevance to the problem of infertility is not known. The experimental use of antibodies against the hormones involved in reproduction raises the question of whether such antibodies are ever of clinical significance in humans.

2. The literature on immunologic significance of cell and protein exchange between mother and fetus is very extensive, especially regarding the levels of bacterial and viral antibody in the mother and infant. Maternal diseases which can be manifest in the infant, because of the placental passage of proteins or antibodies, are reviewed, as are the theoretical aspects of trophoblastemia. The immunologic response of the mother to transferred fetal

platelets, white blood cells, and red blood cells has resulted in thrombocytopenia, agranulocytosis, and hemolytic anemia. Although there is definite evidence of the transfer of fetal cells and antigens to the mother, the extent of the reverse process is not known. From a theoretical standpoint, the transfer of adult competent cells to the fetus might be responsible for the "failure to thrive" syndromes seen in some newborn infants or might be the mechanism whereby autoimmune diseases are initiated.

3. Tissue antibodies are valuable experimental tools for the study of normal and abnormal development in chick and amphibian embryos. Tissue-specific antibodies, when applied directly to the embryo in the proper dose and state, can induce specific alterations just as surgical procedures can result in predictable malformations. Other antibodies produce non-specific alterations and all have the potential for killing the embryo when the dose is high enough.

The experimental production of malformations in mammals is somewhat more complicated since the antibody has the potential for acting on both the embryo and the mother. Furthermore, the interposition of the placenta and fetal membranes introduces a variable that must be evaluated for every experimental or clinical situation. If a generalization is possible at this time, it is that the effects of one antibody in one species cannot be generally applied. Two general lines of investigation have been followed. The more popular approach has been an attempt to produce specific malformations with tissue-specific antibodies. The experimental results include active immunization with brain antigen and thyroid antigen and passive immunization with lens antibody. The results have been equivocal, including specific observations in humans where tissue-specific antibodies have been demonstrated during pregnancy.

On the other hand, tissue antibodies were rarely considered as *general teratogenic agents* in mammals until it was demonstrated that rat-kidney antiserum and rat-placental antiserum are potent teratogenic agents if administered early in gestation. The isolation of a protein produced in mammals with potent teratogenic properties may be significant only to the experimental embryologist or it may be extremely important in the understanding of some aspects of human teratology.

REFERENCES

1. BASKIN, M. J. Temporary sterilization by the injection of human spermatozoa. *Amer. J. Obstet. Gynec.* **24**: 892, 1932.
2. McCARTNEY, J. Mechanism of sterilization of female rat from injections of spermatozoa. *Amer. J. Physiol.* **66**: 404, 1923.
3. POMMERENKE, W. T. Effects of sperm injections into female rabbits. *Physiol. Zool.* **1**: 97, 1928.

4. Freund, J., Lipton, M. M., and Thompson, G. E. Aspermatogenesis in the guinea pig induced by testicular tissue and adjuvants. *J. Exp. Med.* **97**: 711, 1953.

5. Mancini, R. E., Davidson, O. W., Vilar, O., Nemirovsky, M., and Bueno, D. D. Localization of achrosomal antigenicity in guinea pig testes. *Proc. Soc. Exp. Biol. Med.* **111**: 435, 1962.

6. Pokorna, Z., Vojtiskova, M., Rychlikova, M., and Chutna, J. An isologous model of experimental autoimmune aspermatogenesis in mice. *Folia Biologica* (Praha) **9**: 203, 1963.

7. Zuckerman, S. Mechanisms involved in conception. *Science* **130**: 1260, 1959.

8. Boughton, B., and Spector, W. G.: Autoimmune testicular lesions induced by injury to the contralateral testis and intradermal injection of adjuvant. *J. Path. Bact.* **86**: 69, 1963.

9. Vojtiskova, M. and Pokorna, Z. Prevention of experimental allergic aspermatogenesis by thymectomy in adult mice. *Lancet* 644, 1964.

10. Phadke, A. M. and Padukone, K. Presence and significance of autoantibodies against spermatozoa in the blood of men with obstructed vas deferens. *J. Reprod. Fertil.* **7**: 163, 1964.

11. Isojima, S., Graham, R. M., and Graham, J. B. Sterility in female guinea pigs induced by injection with testis. *Science* **129**: 44, 1959.

12. Katsh, S. Infertility in female guinea pigs induced by injection of homologous sperm. *Amer. J. Obstet. Gynec.* **78**: 276, 1959.

13. Katsh, S. Antigenicity of human testis. *J. Urol.* **87**: 896, 1962.

14. Kiddy, C. A., Stone, W. H., and Casida, L. E. Immunologic studies on fertility and sterility. II. Effects of treatment of semen with antibodies on fertility in rabbits. *J. Immun.* **82**: 125, 1959.

15. Otani, Y., and Behrman, S. J. Immunization of the guinea pig with homologous testis and sperm. *Fertil. Steril.* **14**: 456, 1963.

16. Otani, Y., Behrman, S. J., Porter, C. W., and Nakayama, M. Reduction of fertility in immunized guinea pigs. *Int. J. Fertil.* **8**: 835, 1963.

17. Edwards, R. G. Antigenicity of rabbit semen, bull semen, and egg yolk after intravaginal or intramuscular injections into female rabbits. *J. Reprod. Fertil.* **1**: 385, 1960.

18. Freund, J., Lipton, M. M., and Thompson, G. E. Impairment of spermatogenesis in the rat after cutaneous injection of testicular suspension with complete adjuvants. *Proc. Soc. Exp. Biol. Med.* **87**: 408, 1954.

19. Behrman, S. J. and Otani, Y. Transvaginal immunization of the guinea pig with homologous testis and epididymal sperm. *Int. J. Fertil.* **8**: 829, 1963.

20. Franklin, R. and Dukes, C. D. Antispermatozoal antibody and unexplained infertility. *Amer. J. Obstet. Gynec.* **89**: 6, 1964.

21. Rumke, P. H. and Hellinga, G. Auto-antibodies against spermatozoa in sterile men. *Amer. J. Clin. Path.* **32**: 357, 1959.

22. Renkonen, K. O., Makela, O., and Lehtovaara, R. Factors affecting the human sex ratio. *Nature* **194**: 308, 1962.

23. Renkonen, K. O. Decreasing sex-ratio by birth order. *Lancet* **1**: 60, 1963.

24. Edwards, A. W. F. Human sex ratio and maternal immunity to male antigen. *Nature* **198**: 1106, 1963.

25. McLaren, A. Does maternal immunity to male antigen affect the sex ratio of the young? *Nature* **195**: 1323, 1962.

26. Sutherland, J. M. and Landing, B. H. Failure to demonstrate antibody to sperm in serum of prostitutes. *Lancet* **2**: 56, 1961.

27. Collip, J. B. Recent studies on antihormones. *Ann. Int. Med.* **9**: 150, 1935.

28. Ennis, J. E. Immunological pregnancy tests. *Lancet* **2**: 1379, 1963.

29. Yahia, C. and Taymor, M. L. A 3-minute immunologic pregnancy test. *Obstet. Gynec.* **23**: 37, 1964.

30. LEWIS, J., DRAY, S., GENUTH, S., AND SWARTZ, H. Demonstration of immunological similarities of human pregnancy gonadotropin and choriocarcinoma gonadotropin with antisera prepared in rabbits and monkeys. *J. Clin. Endocr.* **24:** 197, 1964.

31. MCGARRY, E. E. AND BECK, J. C. Some studies with antisera to human FSH. *Fertil. Steril.* **14:** 558, 1963.

32. OSBORN, J. J., DANCIS, J., AND JULIA, J. F. Studies of the immunology of the newborn infant. I. Age and antibody production. *Pediatrics* **9:** 736, 1952.

33. OSBORN, J. J., DANCIS, J., AND JULIA, J. F. Studies of the immunology of the newborn infant. II. Interference with active immunization by passive transplacental circulating antibody. *Pediatrics* **10:** 328, 1952.

34. OSBORN, J. J., DANCIS, J., AND ROSENBERG, B. V. Studies of the immunology of the newborn infant. *Pediatrics* **10:** 450, 1963.

35. BRAMBELL, F. W. R., HEMMINGS, W. A., HENDERSON, M., PARRY, H. J., AND ROW-LANDS, W. T. The route of antibodies passing from the maternal to the foetal circulation in rabbits. *Proc. Roy. Soc. (Biol.)* **136:** 131, 1949.

36. BRAMBELL, F. W. R. The passive immunity of the young mammal. *Biol. Rev.* **33:** 488, 1958.

37. DANCIS, J., LIND, J., ORATZ, M., SMOLENS, J., AND VERA, P. Placental transfer of proteins in human gestation. *Amer. J. Obstet. Gynec.* **82:** 167, 1961.

38. LIEBLING, J. AND SCHMITZ, H. E. Protection of infant against diphtheria during first year of life following active immunization of pregnant mother. *J. Pediat.* **23:** 430, 1943.

39. BANGHAM, D. R., HOBBS, K. R., AND TEE, D. E. H. Transmission of serum proteins from fetus to mother in the rhesus monkey. *Lancet* **2:** 1173, 1960.

40. IWAMOTO, K. Experimental studies on the fetal transfer of antigen introduced into the pregnant rabbit, with particular reference to the pathway for the antigen transfer. *Acta Med. Biol.* **8:** 29, 1960.

41. GITLIN, D., ROSEN, F. S., AND MICHAEL, J. G. Transient 19 S gamma$_1$ globulin deficiency in the newborn infant and its significance. *Pediatrics* **31:** 197, 1963.

42. FREDA, V. J. AND CARTER, B. Placental permeability for the beta isoantibodies in relation to the maternal A_1–A_2 subgroups. *Obstet. Gynec.* **17:** 597, 1961.

43. BERLYNE, G. M., SHORT, I. A., AND VICKERS, C. F. H. Placental transmission of the L.E. factor. *Lancet* **2:** 15, 1957.

44. HOGG, G. R. Congenital, acute lupus erythematosus associated with subendocardial fibroelastosis. *Amer. J. Clin. Path.* **28:** 648, 1957.

45. SEIP, M. Systemic lupus erythematosus in pregnancy with haemolytic anaemia, leucopenia, and thrombocytopenia in the mother and her newborn infant. *Arch. Dis. Child.* **35:** 364, 1960.

46. JACKSON, R. Discoid lupus in a newborn infant of a mother with lupus erythematosus. *Pediatrics* **33:** 425, 1964.

47. EPSTEIN, R. D., LOZNER, E. L., COBBEY, T. S., AND DAVIDSON, C. S. Congenital thrombocytopenic purpura. *Amer. J. Med.* **9:** 44, 1950.

48. SLOBODY, L. B., ABRAMSON, H., AND LOIZEAUX, L. S. Agranulocytosis of the newborn infant. *J. Amer. Med. Ass.* **142:** 25, 1950.

49. LUHBY, A. L. AND SLOBODY, L. B. Transient neonatal agranulocytosis in two siblings. Transplacental immunization to a leukocyte factor. *Amer. J. Dis. Child.* **92:** 496, 1956.

50. PAYNE, R. AND ROLFS, M. R. Fetomaternal leukocyte incompatibility. *J. Clin. Invest.* **37:** 1756, 1958.

51. NAESLUND, J. AND NYLIN, G. Investigation on the permeability of the placenta with the aid of red blood corpuscles tagged with radioactive phosphorus. *Acta Med. Scand.* **170:** 390, 1946.

52. CHOWN, B. The fetus can bleed. *Amer. J. Obstet. Gynec.* **70:** 1298, 1955.

53. ZIPURSKY, A., HULL, A., WHITE, F. D., AND ISRAELS, L. G. Fetal erythrocytes in the maternal circulation. *Lancet* **1:** 451, 1959.

Robert L. Brent

54. BROWN, W. E. AND COWELS, G. T. Fetal cells in maternal blood. *Southern Med. J.* **56**: 782, 1963.

55. THOMAS, L., DOUGLAS, G. W., AND CARR, M. C. The continual migration of syncytial trophoblasts from the fetal placenta into the maternal circulation. *Ass. Amer. Physicians* **72**: 140, 1959.

56. RUGGIERI, P., BOLOGNESI, G., GATTA, L., AND MARCHI, M. Dimostrazione di antigeni placentari nel siero di donne gravida. *Progr. Med. Nap.* **14**: 449, 1958.

57. ESPOSITO, A. Ricerche immunoelectroforetiche nel siero di donne gravide. *Arch. Ostet. Ginec.* **65**: 818, 1960.

58. SIMMONS, R. L. AND RUSSELL, P. S. The antigenicity of mouse trophoblast. *Ann. N.Y. Acad. Sci.* **99**: 717, 1962.

59. RETIK, A. B., SABESIN, S. M., HUME, R., MALGREN, R. A., AND KETCHAM, A. S. The experimental transmission of malignant melanoma cells through the placenta. *Surg. Gynec. Obstet.* **114**: 485, 1962.

60. DIAMANDOPOULOS, G. T. AND HERTIG, A. T. Transmission of leukemia and allied diseases from mother to fetus. *Obstet. Gynec.* **21**: 150, 1963.

61. HALASZ, N. A. AND ORLOFF, M. Transplacental transmission of homotransplantation antibodies. *J. Exp. Med.* **118**: 353, 1963.

62. GUYER, N. F. AND SMITH, E. A. Studies on cytolysins. I. Some prenatal effects of lens antibodies. *J. Exp. Zool.* **26**: 65, 1918.

63. GUYER, M. F. AND SMITH, E. A. Studies on cytolysins. II. Transmission of induced eye defects. *J. Exp. Zool.* **31**: 171, 1920.

64. FINLAY, G. F. The effect of different species lens antisera on pregnant mice and rats and their progeny. *Brit. J. Exp. Biol.* **1**: 201, 1924.

65. HUXLEY, J. S. AND CARR-SAUNDERS, A. M. The absence of prenatal effects of lens-antisera in rabbits. *Brit. J. Exp. Biol.* **1**: 215, 1924.

66. POYNTER, C. W. M. AND ALLEN, E. V. Lens antigen as a factor in congenital and hereditary eye abnormalities. *Amer. J. Ophthal.* **8**: 184, 1925.

67. DAVIS, F. A. AND SMITH, H. M. The role of lens antigen and uveal pigment in the production of hereditary anomalies of the eye. *Arch. Ophthal.* **4**: 672, 1930.

68. DAVIS, F. A. Hereditary eye defects in rabbits experimentally induced. *Trans. Ophthal. Soc.* **45**: 555, 1926.

69. MILLER, W. J. Anti-lens sera as a mutagen in rabbits. *J. Exp. Zool.* **137**: 463, 1958.

70. FOWLER, I. AND CLARKE, W. M. Development of anterior structures in the chick after direct application of adult lens antisera. *Anat. Rec.* **136**: 194, 1960.

71. CLARKE, W. M. AND FOWLER, I. The inhibition of lens-inducing capacity of the optic vesicle with adult lens antisera. *Develop. Biol.* **2**: 155, 1960.

72. LANGMAN, J. The effect of lens antiserum on chick embryos. *Anat. Rec.* **137**: 135, 1960.

73. LANGMAN, J. The effects of antibodies on embryonic cells. *Canad. Cancer Conf.* **5**: 349, 1963.

74. EBERT, J. An analysis of the effects of anti-organ sera of the development, *in vitro*, of early chick blastoderm. *J. Exp. Zool.* **115**: 351, 1950.

75. JOHNSON, I. S. AND LEONE, C. A. The ontogeny of proteins of the adult chicken heart as revealed by serological techniques. *J. Exp. Zool.* **130**: 515, 1955.

76. CLAYTON, R. M. Distribution of antigens in the developing newt embryo. *J. Embryol. Exp. Morph.* **1**: 25, 1953.

77. NACE, G. W. AND INOUE, K. Cytolysis versus differentiation in antineurula serum. *Science* **126**: 259, 1957.

78. LENICQUE, P. Studies on homologous inhibition in the chick embryo. *Act. Zool. Stockh.* **40**: 141, 1959.

79. CLARKE, R. B. AND McCALLION, D. J. Specific inhibition of neuronal differentiation in the chick embryo. *Canad. J. Zool.* **37**: 133, 1959.

80. McCALLION, D. J. AND LANGMAN, J. An immunologic study on the effect of brain extract on the developing nervous tissue in the chick embryo. *J. Embryol. Exp. Morph.* **12**: 77, 1964.

81. SACERDOTE DE LUSTIG, E. AND FISZER, B. Action del gamma globuline du serum humain normal ou cancereux sur l'embryon de Poulet. *C.R. Soc. Biol.* **150**: 1795, 1956.

82. SACERDOTE DE LUSTIG, E. AND FISZER, B. Accion de la fraccion gamma globulina del suero canceroso sobre el embrion de pollo. *Rev. Soc. Argent. Biol.* **32**: 120, 1957.

83. CHOLEWA, L. Wpxyw surowicy chorych na raka na rozwoj kurzych zarodkow. *Polski Tygodnik Lekarski* **12**: 929, 1957.

84. LACON, C. R. AND KARNOFSKY, D. Effects of blood taken from patients with neoplastic disease on the chick embryo. *Proc. Soc. Exp. Biol. Med.* **115**: 477, 1964.

85. LANGMAN, J., VAN DRUNEN, H., AND BOWMAN, F. Maternal protein metabolism and embryonic development in human beings. *Amer. J. Obstet.* **77**: 549, 1959.

86. GILLMAN, J., GILBERT, C., GILLMAN, T., AND SPENCE, L. A preliminary report on hydrocephalus, spina bifida and other congenital anomalies in the rat produced by trypan blue. *S. Afr. J. Med. Sci.* **13**: 47, 1948.

87. DIJKSTRA, J. AND GILLMAN, J. Trypan blue concentration and protein composition in sera of rats injected repeatedly with trypan blue in relation to reticulosis and to reticulosarcoma. *S. African J. Med. Sci.* **25**: 119, 1960.

88. BEAUDOIN, A. R. AND KAHKONEN, D. The effect of trypan blue on serum proteins of the fetal rat. *Anat. Rec.* **147**: 387, 1963.

89. OCHIAC, Y. Allergic histological changes in the pregnant rabbit organs induced by injection of the globulin fractions from human placenta. 2. Histological studies of the liver of rabbit treated with protein fraction from normal and toxemic human placenta. *Acta Med. Okayama* **14**: 279, 1960.

90. LAW, R. R. AND LAW, R. Abortion after gamma globulin. *Brit. Med. J.* **5359**: 747, 1963.

91. SEEGAL, B. C. AND LOEB, E. N. Effect of antiplacenta serum on development of the fetus in the pregnant rat. *Proc. Soc. Exp. Biol. Med.* **45**: 248, 1940.

92. LOEB, E. N. AND SEEGAL, B. D. The production of chronic nephritis in the rat following the initial injection of anti-placenta serum. II. Pathological findings. *Fed. Proc.* **2**: 1943.

93. SEEGAL, B. C. AND LOEB, E. N. Production of chronic glomerulonephritis in rats by injection of rabbit antirat-placenta serum. *J. Exp. Med.* **84**: 211, 1946.

94. LOEB, E. M., KNOWLTON, A. I., STOERK, H. D., AND SEEGAL, B. C. Observations of the pregnant rat injected with nephrotoxic rabbit antirat-placenta serum and desoxycorticosterone acetate. *J. Exp. Med.* **89**: 287, 1949.

95. BEVANS, M., SEEGAL, B. C., AND KAPLAN, R. Glomerulonephritis produced in dogs by specific antisera. II. Pathologic sequences following the injection of rabbit antidog-placenta serum or rabbit antidog-kidney serum. *J. Exp. Med.* **102**: 807, 1955.

96. SEEGAL, B. C., HANSSON, M. W., GAYNOR, E. C., AND ROTHENBERG, M. C. Glomerulonephritis produced in dogs by specific antisera. I. The course of the disease resulting from injection of rabbit antidog-placenta serum or rabbit antidog-kidney serum. *J. Exp. Med.* **102**: 789, 1955.

97. McCAUGHEY, W. T. E. The nephrotoxic action of antiplacenta serum in rats. *J. Obstet. Gynaec. Brit. Emp.* **62**: 863, 1955.

98. OLIVELLI, F. Sull'azione dei sieri antiplacentari nella cavia gravida. *Minerva Ginec.* **10**: 131, 1958.

99. PRESSMAN, D. AND KORNGOLD, L. Localizing properties of anti-placenta serum. *J. Immun.* **78**: 75, 1957.

100. STEBLAY, R. W. Closely related antigens in human placenta and kidney. *Nature* **192**: 1259, 1961.

101. STEBLAY, R. W. Localization in human kidney of antibodies formed in sheep against human placenta. *J. Immun.* **88:** 434, 1962.

102. STEBLAY, R. W. Effect of injecting heterologous glomerular basement membrane preparations in pregnant sheep. *Proc. Soc. Exp. Biol. Med.* **112:** 15, 1963.

103. GLUECKSOHN-WAELSCH, S. The effect of maternal immunization against organ tissues on embryonic differentiation in the mouse. *J. Embryol. Exp. Morph.* **5:** 83, 1957.

104. BARBER, A. N., WILLIS, J., AND AFEMAN, C. Changes in the lens induced by maternal hypersensitivity in mice. *Amer. J. Ophthal.* **51:** 949, 1961.

105. LEVI-MONTALCINI, R. Nerve-growth controlling factors. First International Conference on Congenital Malformations. London, 1961. Philadelphia: Lippincott.

106. BRENT, R. L., AVERICH, E., AND DRAPIEWSKI, V. A. Production of congenital malformations using tissue antibodies. I. Kidney antisera. *Proc. Soc. Exp. Biol. Med.* **106:** 523, 1961.

107. DAVID, G., MERCIER-PAROT, L., AND TUCHMANN-DUPLESSIS, H. Action tératogène d'hetero-anticorps tissulaires. I. Production de malformations chez le rat par action d'un serum anti-rein. *Comptes rendus des seances de la Société de Biologie* **157:** 939, 1963.

108. BRENT, R. L. The production of congenital malformations with tissue antibodies. II. The spectrum and incidence of malformations following the administration of kidney antiserum to the pregnant rat. *Fed. Proc.* **23:** 390, 1964.

109. BRENT, R. L. The production of congenital malformations using tissue antisera. II. The spectrum and incidence of malformations following the administration of kidney antiserum to pregnant rats. *Amer. J. Anat.* **115:** 1964.

110. BRENT, R. L. The production of congenital malformations using tissue antisera. III. Placenta antiserum. American Pediatric Society, 1964 (abstr.).

111. BRENT, R. L. AND ZIEGRA, S. R. Effect of teratogenic kidney-antibody on placenta grown in tissue culture. *Amer. J. Dis. Child.* **104:** 464, 1962.

112. BLIZZARD, R. M., CHANDLER, R. W., LANDING, B. H., PETTIT, M. D., AND WEST, C. D. Maternal autoimmunization to thyroid as a probable cause of athyrotic cretinism. *New Engl. J. Med.* **263:** 327, 1960.

113. SUTHERLAND, J. M., ESSELBORN, V. M., BURKET, R. L., SKILLMAN, T. B., AND BENSON, J. T. Familial non-goitrous cretinism apparently due to maternal antithyroid antibody. *New Engl. J. Med.* **263:** 336, 1960.

114. CHANDLER, R. W., KYLE, M. A., HUNG, W., AND BLIZZARD, R. M. Experimentally induced autoimmunization disease of the thyroid. I. The failure of transplacental transfer of anti-thyroid antibodies to produce cretinism. *Pediatrics* **29:** 961, 1962.

115. HELLWIG, C. A. AND WILKINSON, P. N. Experimental production of chronic thyroiditis. *Growth* **24:** 169, 1960.

116. HUNG, W., MIGEON, C., AND PARROT, R. H. A possible autoimmune basis for Addison's disease in three siblings, one with idiopathic hypoparathyroidism, pernicious anemia and superficial moniliasis. *New Engl. J. Med.* **269:** 658, 1963.

UTERINE VASCULAR CLAMPING AND OTHER SURGICAL TECHNIQUES IN EXPERIMENTAL MAMMALIAN EMBRYOLOGY

I. Demonstration of the technique of uterine vascular clamping in the pregnant rat.

  A. Discussion of anesthesia techniques

B. Demonstration of clamping procedure, types of clamps, recording of data, and methods of closing the incision

C. Discussion of the results of uterine vascular clamping and the application of these findings in the fields of fetal pharmacology and experimental embryology

D. Movie and slides depicting various surgical techniques and procedures

II. Immunology

Often when describing a new test or procedure, the author leaves unsaid many of the pertinent facts necessary to complete that procedure satisfactorily. The purpose of this portion of the laboratory session is to describe some of the frequently overlooked minor points which make the difference between unsatisfactory and satisfactory results.

   A. Agar diffusion plates
      1. Cut-outs, patterns, and templates
      2. Methods of preparation
      3. Interpretation
      4. Photography
   B. Protein separation and fractionation techniques
   C. Fluorescence labeling (time permitting)
   D. Demonstration of the effect of rabbit anti-rat kidney antibody in a pregnant rat

III. Equipment
   A. Restraining animal board
   B. Restraining clamps or adhesive tape
   C. Razor blades, surgical scissors, hemostat, needle holder, forceps
   D. Sodium pentobarbital (10 mg/ml)
   E. Saline (0.85 per cent)
   F. Surgical sponges ($4'' \times 4''$ and $3'' \times 3''$)
   G. Silk suture (000) and cutting-edge needle
   H. Tracheal suction apparatus
   I. Wound clips and applicator
   J. Alcohol and cotton sponges
   K. Abdominal retractors
   L. Biliary clamps with longitudinal serrations

OPERATIVE TECHNIQUES INCLUDING THE UTERINE VASCULAR CLAMPING PROCEDURE

Obtaining adequate anesthesia in pregnant animals is difficult because not only is one concerned with immobilization and analgesia in the adult organism but also assurance that the anesthesia does not alter the status or environment of the embryo or fetus. Inhalation anesthesia has proved to be

Robert L. Brent   234

deleterious to the fetus under laboratory conditions because of difficulty in controlling the concentration of ether or chloroform that the mother receives. Each animal may receive a different dose of anesthesia for a different length of time resulting in a spectrum of levels of anesthesia. This leads to unpredictable incidences of whole-litter resorptions in the mouse and rat.

We have found sodium pentobarbital to be the most satisfactory drug to use in the rat, mouse, and rabbit. The dose is based on the weight of the pregnant animal on the day of operation and is 30 mg/kg for the rat and 90 mg/kg for the mouse. This dose will afford 2.5 hours of light anesthesia in the rat and 1.5 hours in the mouse if given subcutaneously. Peritoneal injection will result in slightly deeper and shorter periods of anesthesia. The length and depth of anesthesia can also be controlled by the volume of the injected solution. (Veterinary pentobarbital contains 60 mg/ml; we dilute it to 10 mg/ml.) During anesthesia, secretions may accumulate in the pharynx which can be aspirated with a 10-ml syringe attached to a polyethylene tube (P.E. 90, Clay-Adams). Gentle suction will remove these secretions and prevent sudden death in these animals from tracheal obstruction.

Because of the vasodilatory properties of pentobarbital, it is important to keep the animals warm during and after surgery. A 100-watt lamp, properly placed, is adequate for this purpose.

For any abdominal operative procedure, we shave the abdomen and cleanse the area with alcohol. The surgical instruments are clean, although not necessarily sterile. The silk suture material (000) is sterilized. If vials of synthetic sutures are utilized, they must be soaked for some time to remove a preservative which can produce sterile abscesses in the abdominal wall. The musculature is closed with interrupted sutures, and the skin is closed with wound clips.

In manipulating the viscera, blunt, flat instruments should be used. Before removing the uterus or other viscera on the abdominal wall, saline-soaked sponges should cover the field and, if the organs are to be removed from the peritoneal cavity for any length of time, they should also be covered with saline-soaked sponges.

The clamping procedure is quite simple (Plate 1). Curved gallbladder clamps are placed across the uterus and its mesentery at the cervical and ovarian end of one or both horns in the rat. The clamps should rest on the operative board so as not to produce undue tension or pressure on any of the viscera. After a predetermined period of time (see Table 1), the clamps are removed, the viscera returned to the abdominal cavity, and the incision closed.

Other surgical techniques that can be carried out similarly are (1) mobilization of the uterus by severing the ovarian artery (Plates 2–4); (2) transillumination of the uterus to locate the position of the placenta (Plate 5,

TABLE 1

EFFECT OF UTERINE VASCULAR CLAMPING ON EMBRYONIC MORTALITY AND GROWTH

Time Clamped	Embryos in the Control (Unclamped) Uterine Horn				Embryos in the Clamped Uterine Horn			
	Control Embryos	Per Cent Mortality	Term No.	Fetuses (Weight and SD)	Clamped Embryos	Per Cent Mortality	Term No.	Fetuses (Weight and SD)
3-Day Embryos								
½ hr........	22	9.1	20	5.06±0.47	....	Very high	4	5.44±0.41
1 hr........	27	3.7	26	4.98±0.58	....	100	0	....
4-Day Embryos								
½ hr.......	48	33.3	32	4.93±0.65	43	81.4	8	4.66±0.72
1 hr.......	37	24.3	28	4.84±0.64	38	71.0	11	4.73±0.67
5-Day Embryos								
½ hr.......	26	50.0	13	5.28±0.54	32	75.0	8	4.79±0.47
1 hr.......	41	24.3	31	4.93± .45	24	71.8	7	4.81± .60
1½ hr......	22	18.1	18	4.87± .40	29	62.0	11	4.02±0.38
2 hr.......	37	32.4	25	4.51±0.65	22	90.9	2	3.84
6-Day Embryos								
½ hr.......	21	0	21	5.38±0.56	35	20.0	28	5.17±0.55
1 hr.......	13	23.1	10	5.19± .33	31	32.5	21	4.59± .89
1½ hr......	22	4.5	21	5.04± .41	32	31.3	22	4.54± .70
2 hr.......	14	7.1	13	5.21± .30	29	79.3	6	4.57± .32
2½ hr......	17	11.8	15	5.14±0.55	32	90.6	3	4.89±0.10
7-Day Embryos								
½ hr.......	27	7.4	25	4.59±0.50	34	5.8	32.	4.59±0.24
1 hr.......	20	15.0	17	4.82± .17	41	17.0	34	4.54± .46
1½ hr......	15	20.0	12	5.08± .63	34	64.7	12	4.71± .80
2 hr.......	31	35.4	20	5.20± .32	42	64.2	15	4.10± .99
2½ hr......	16	18.7	13	4.86±0.71	31	67.7	10	4.62±0.31
8-Day Embryos								
½ hr.......	45	7.1	42	5.03±0.43	75	21.3	59	4.59±0.54
1 hr.......	30	10.0	27	4.58± .63	36	38.9	22	3.99±0.45
1½ hr......	.....	.....	.....	.....	.....	.....	.....	.....
2 hr.......	45	24.4	34	4.57	67	53.7	31	3.50±1.16
2½ hr......	33	15.1	28	5.02± .57	54	68.5	17	3.77±0.71
3 hr........	25	16.0	21	4.77±0.71	38	89.5	4	3.33±0.59

TABLE 1—*Continued*

Time Clamped	Embryos in the Control (Unclamped) Uterine Horn				Embryos in the Clamped Uterine Horn			
	Control Embryos	Per Cent Mortality	Term No.	Fetuses (Weight and SD)	Clamped Embryos	Per Cent Mortality	Term No.	Fetuses (Weight and SD)
				9-Day Embryos				
½ hr.......	25	12.0	22	4.54±0.55	33	15.2	28	4.42±0.58
1 hr.......	21	4.8	20	4.91± .36	29	27.6	21	4.57± .46
1½ hr.......	24	8.4	22	4.90± .32	39	41.1	23	4.50± .63
2 hr.......	49	14.3	42	4.94± .35	70	88.6	8	3.83± .84
2½ hr.......	25	24.0	19	4.90± .77	34	50.0	17	4.20±0.64
3 hr.......	18	27.8	13	4.50±0.86	21	95.2	1	....
				10-Day Embryos				
½ hr.......	24	16.7	20	4.48±0.35	38	26.3	28	4.43±0.41
1 hr.......	23	13.0	20	4.69±0.84	38	26.3	28	4.25± .64
1½ hr.......	10	20.0	8	4.53±1.24	31	25.8	23	4.49± .66
2 hr.......	48	20.8	38	4.81±0.33	68	67.6	22	4.26± .66
2½ hr.......	21	19.0	17	4.29±0.46	33	84.8	5	3.63±0.38
				11-Day Embryos				
½ hr.......	19	26.3	14	4.88±0.41	41	19.5	33	4.71±0.43
1 hr.......	20	0	20	4.73± .38	35	40.0	21	4.28± .44
1½ hr.......	22	13.6	19	4.76± .30	29	41.4	17	4.46± .49
2 hr.......	24	8.3	22	4.69± .45	42	66.7	14	4.42± .49
2½ hr.......	28	0	28	4.79±0.40	32	56.2	14	4.70±0.53
				12-Day Embryos				
½ hr.......	30	40.0	18	4.61±0.32	41	43.9	23	4.55±0.81
1 hr.......	18	0	18	4.75± .32	32	25.0	24	4.55± .47
1½ hr.......	20	17.6	17	4.97± .46	35	85.7	5	4.93±0.71
2 hr.......	15	6.7	14	5.07±0.33	33	100	0	....
				13-Day Embryos				
½ hr.......	25	20.0	20	4.43±0.58	31	35.5	20	4.64±0.46
1 hr.......	33	48.5	17	5.08± .47	39	74.4	10	4.85±0.47
1½ hr.......	17	41.2	10	4.87±0.48	37	100	0	....
				14-Day Embryos				
½ hr.......	24	45.8	13	4.92±0.60	33	48.5	17	4.58±0.60
1 hr.......	26	53.8	12	4.79± .40	36	61.1	14	4.81± .45
1½ hr.......	20	50.0	10	4.51±0.20	40	82.5	7	4.19±0.57
2 hr.......	26	73.1	7	....	34	82.4	6	....

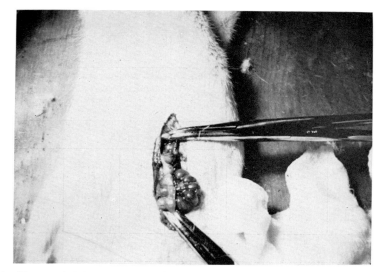

PLATE 1.—Two special clamps are placed at the ovarian and cervical ends of one or both uterine horns, across the uterus and blood vessels in the mesentery. Thus, the ovarian artery and vein and the uterine artery and vein are clamped. In this way the uterine horn is completely separated from the maternal circulation. The clamping procedure has been carried out for from 15 minutes to 3 hours. (See Table 1 and Selected Bibliography for applications.)

PLATE 2.—The ovarian artery and vein can be surgically severed early in gestation or during the period of differentiation to give greater mobility to the uterus for exteriorizing procedures without interfering with the development of the embryos. A bulldog clamp has been placed across the ovarian stump and the pregnant uterus is lying on the abdominal wall. The uterus can just as readily be mobilized at the cervical end but, of course, this will eliminate a natural delivery, and the fetus will have to be delivered by caesarian section.

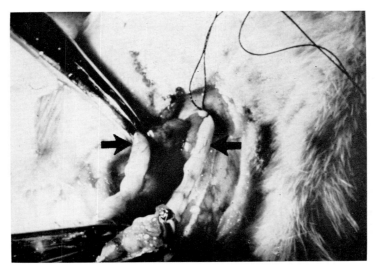

PLATE 3.—An 8-day-pregnant rat uterus has been mobilized at the ovarian end on the animal's left. The suture material is allowed to remain long for ease in manipulating the uterus. Both cervical ends of the two uterine horns (*arrows*) have been clamped as has the ovarian end of the right uterine horn. This is in preparation for external oxygenation of the left uterine horn in combination with bilateral uterine vascular clamping as depicted in Plate 4.

PLATE 4.—The left uterine horn has been placed in a temperature-controlled oxygen chamber. The large arrow points to the temperature-control chamber; the small arrow points to the inner chamber containing the uterus. This view was made possible by severing the ovarian artery and vein and mobilizing the uterus. This procedure demonstrates the extent of the manipulations that the pregnant rat uterus will tolerate.

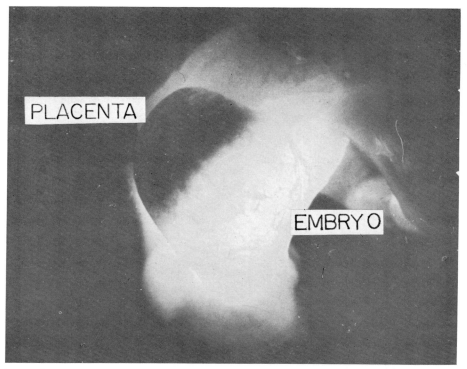

PLATE 5.—The placenta and embryo can be identified by transilluminating the pregnant rat uterus, providing the embryo is 12 days old or older.

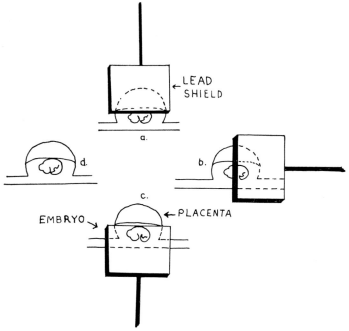

FIG. 1.—Once the line of demarcation between the placenta and embryo is identified, various procedures can be carried out, including differential irradiation or injecting substances into the embryo or placenta. This figure depicts shielding techniques for irradiating the embryo or placenta separately.

Fig. 1); (*3*) differential irradiation and shielding procedures using X-ray (Plates 6–10, Fig. 3); (*4*) temperature control of the exteriorized uterus.

Good record keeping is critical to any embryological experiment. The record sheet we utilize is versatile enough for our needs, but each investigator should design his own record sheet, depending on his experimental goals. We redesign our record sheet periodically to fill the changing needs of the laboratory (Figs. 3 and 4).

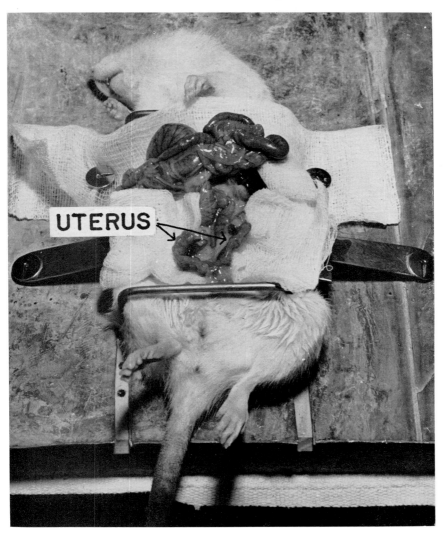

PLATE 6.—Plates 6–8 depict three operative stages of mobilizing and shielding the pregnant rat uterus. In this manner about 90 per cent of the mother is irradiated and the embryos receive practically no irradiation. The lead shields are coated with a thin layer of paraffin to protect the blood vessels and tissues that they contact.

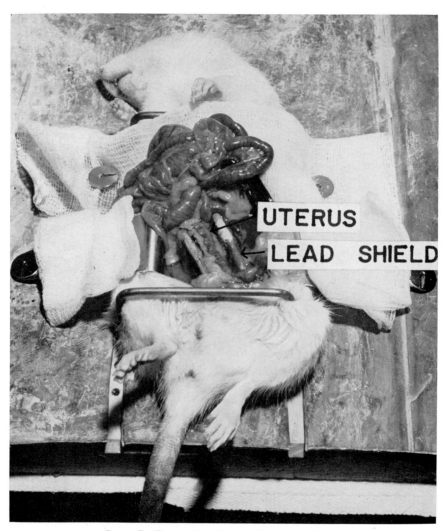

PLATE 7.—For explanation, see legend to Plate 6.

PLATE 8.—For explanation, see legend to Plate 6.

PLATE 10

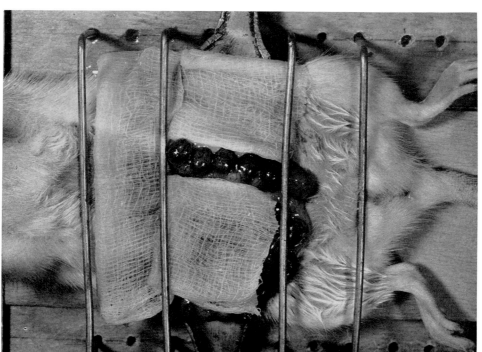

PLATE 9

PLATES 9 AND 10.—These two plates depict the procedure for irradiating only a portion of the embryos in a pregnant animal. Thus, multiple dosage groups and controls can be obtained from the same litter. The uterus is fixed by gently packing gauze sponges on either side. Wire wickets are inserted into the operative board to support the weight of the lead shields.

244

The pregnant rat uterus is obviously very resistant to surgical procedures. Because of this resistance and the bicornuate condition of the uterus extensive operative procedures can be performed on one uterine horn while embryos in the other horn can be left untreated to serve as littermate controls. Figures 1 and 2 and Plates 1–10 depict some of the surgical and manipulative procedures that have been performed in our laboratory. One word of caution is necessary. The average number of implantation sites in the left and right horn are not the same and, therefore, if one uses one horn

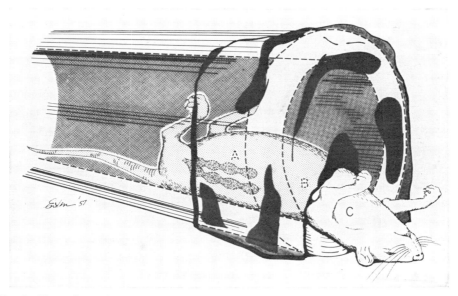

Fig. 2.—If one is not interested in irradiating the greater portion of the mother but wishes to deliver the same gram-roentgen dose, then just the head portion of the maternal organism can be irradiated. This is a much simpler technique, but it offers some theoretical disadvantages. The pregnant rat's abdomen is placed in a lead cylinder and the thorax and head dropped over the edge. Another lead shield with a cutout the shape of the thorax is placed over the opening in the lead cylinder.

consistently as the experimental side, he is bound to obtain unreal "experimental" differences between the control and experimental sides (Fig. 5).

Further applications of these procedures are described in the papers in the selected bibliography and include the use of uterine vascular clamping as an adjunct in drug testing.

IMMUNOLOGY AND IMMUNODIFFUSION SYSTEMS

Since there are excellent reference works in the field of immunology dealing with the experimental techniques used in immunology, it would be a wasted effort to duplicate this material since we are mainly considering

YEAR   TENS   UNITS   EXPERIMENT NO._____

EAR MARK_____

2   4   4   2

1   1

Gestational Stage of Experiment_____ Sacrifice_____

Left Side Clamped_____ Initiated_____ Released_____

Right Side Clamped_____ Initiated_____ Released_____

Agent Administered_____ Route_____ Time_____

        Dose_____mg/kg_____ml_____conc.

Temperature Control_____

_____

Other_____

_____

_____Many _____Few Sperm

AGE	DATE	WEIGHT
0		
1		
2		
3		
4		
5		
6		
7		
8		
9		
10		
11		
12		
13		
14		
15		
16		
17		
18		
19		
20		
21		

DAY OF OPERATION   DAY OF SACRIFICE

Place embryonic sites and ovaries in relative positions.

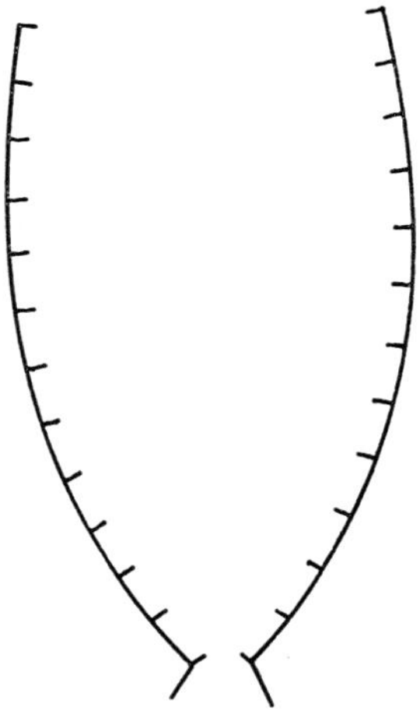

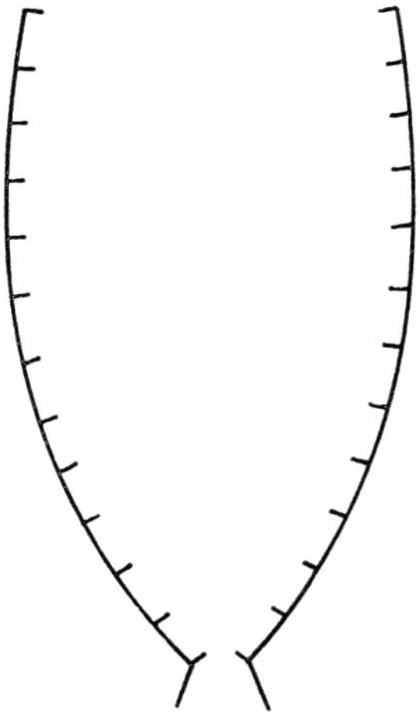

FIG. 3.—Record sheet for each inseminated female that is participating in an experiment.

STATISTICS

	Corp. Lut. Term	Imp. Exp. Day	Resorp.	% Mort.	Term Fetus No.(n)	Weight			
						Sx	Sx^2	(Sx)^2	$\bar{x}$
Left									
Right									
Total									

Left Placental Wt._____ No._____ Av._____

Right Placental Wt._____ No._____ Av._____

Total Placental Wt._____ No._____ Av._____

Uristix Urine Protein_____mg/100 ml_____

Fluorometer Urine Protein_____ g %_____

Fluorometer Serum Protein_____ g %_____

_____

_____

_____

_____

	WEIGHT	COMMENT
A		
B		
C		
D		
E		
F		
G		
H		
I		
J		
K		
L		
M		
N		
O		
P		
Q		
R		

Fig. 4.—Reverse side of record sheet for additional information.

247

embryological techniques. The important references are listed in the Selected Bibliography.

Sephadex reference cards 1–164 are available from Pharmacia Fine Chemicals, Inc., 501 Fifth Avenue, New York 17, N.Y.

Recent references on fluorescent labeling and protein separation are also in the bibliography.

We would like to describe our experience with agar diffusion techniques

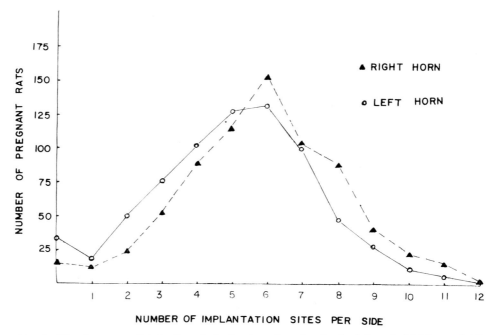

Fig. 5.—The frequency distribution of the implantation sites in the left and right horn vividly demonstrates that the distribution is not the same in the Wistar rat. Therefore, it is important in all experiments in which control embryos are obtained from the same litters as experimental embryos to alternate the side of the experimental procedures. In our rat colony, the mean number implantations in the left uterine horn was 5.0, while the right uterine horn had an average of 5.5 implantation sites.

to emphasize the importance of detail in obtaining reproducible results. The concentration of agar is critical and should be constant from one plate to another. We use 1 per cent agar. Because of the length of time it takes for these plates to develop, a preservative should be added to the agar to prevent bacterial and fungal growth (0.5 per cent phenol).

The following factors should be held constant if one wishes to be able to compare a reaction in one plate with a precipitin pattern in another plate:

1. Thickness of layer of agar.
2. Temperature.
3. Volume of antigen and antibody solutions (and knowledge of their concentration, when possible).
4. Size and distance between wells.

The best way to accomplish these goals is to use lucite templates with exactly the same pattern. This system eliminates any holes in the agar. The distance between the wells is 0.5 inch, but this is dependent on the antigen and antibody concentration and the size of the well and must be individualized.

After the plates have developed, the plastic template is removed, the plate is flooded with distilled water, and it is photographed through the water surface. The key points in photography are Kohler illumination of the plates at a 20° angle from the vertical, high contrast film and printing paper, and absolute darkness above the plate at the time of photographing.

SELECTED BIBLIOGRAPHY

BRAMBELL, W. F. R., HEMINGS, W. A., AND HENDERSON, M. *Antibodies and embryos.* London: Athlone Press, 1951.

BRENT, R. L. The indirect effect of irradiation on embryonic development. II. Irradiation of the placenta. *Amer. J. Dis. Child.* **100:** 103, 1960.

BRENT, R. L. "Modification of teratogenic and lethal effects of irradiation to the mammalian fetus." W. D. Carlson, ed. New York: Pergamon Press, 1963.

BRENT, R. L. AND FRANKLIN, J. B. Uterine vascular clamping: New procedures for the study of congenital malformations. *Science* **132:** 89, 1960.

BRENT, R. L. AND MCLAUGHLIN, M. M. The indirect effect of irradiation on embryonic development. I. Irradiation of the mother while shielding the embryonic site. *Amer. J. Dis. Child.* **100:** 94, 1960.

BRENT, R. L., BOLDEN, B. T., AND FRANKLIN, J. B. Evaluation of teratogenic agents by means of the uterine vascular clamping technique. *Amer. J. Dis. Child.* **104:** 464, 1962.

BRENT, R. L., FRANKLIN, J. B., AND BOLDEN, B. T. The reduction of fetal mortality and malformation from X-irradiation by the use of the uterine vascular clamping technique. *Fed. Proc.* **20** (part I): 398, 1961 (abstr.).

BRENT, R. L., FRANKLIN, J. B., AND BOLDEN, B. T. Modifications of irradiation effects on rat embryos by uterine vascular clamping. *Radiat. Res.* **118:** 58, 1963.

BRENT, R. L., FRANKLIN, J. B., GOLDFARB, A. F., AND MATSUMOTO, R. Modification by progestational compounds of the teratogenic effect of uterine vascular clamping. *Fertil. Steril.* **14:** 365, 1963.

CHERRY, W. B., GOLDMAN, M., CARSKI, T. R., AND MOODY, M. D. Fluorescent antibody techniques. Public Health Service Publication No. 729. U.S. Department of Health, Education, and Welfare, Public Health Service, 1960.

CROWLE, A. J. *Immunodiffusion.* New York and London: Academic Press, 1961, p. 333.

FELTON, L. C. AND MCMILLION, C. R. Chromatographically pure fluorescein and tetramethylrhodamine isothiocyanates. *Anal. Biochem.* **2:** 178, 1961.

FRANKLIN, J. B. AND BRENT, R. L. Interruption of uterine blood flow: A new technique for the production of congenital malformations: Comparison of the eighth, ninth, and tenth days of gestation. *Surg. Forum* **11:** 415, 1960.

FRANKLIN, J. B. AND BRENT, R. L. The effect of uterine vascular clamping on the development of rat embryos 3 to 14 days old. *J. Morph.* **115:** 273, 1964.

GOLDSTEIN, G., SLIZYS, I. S., AND CHASE, M. W. Studies on fluorescent antibody staining. I. Non-specific fluorescence with fluorescein-coupled sheep antirabbit globulins. *J. Exp. Med.* **114:** 89, 1961.

GOLDSTEIN, G., SPALDING, B. H., AND HUNT, W. B. Studies on fluorescent antibody staining II. Inhibition by sub-optimally conjugated antibody globulins. *Proc. Soc. Exp. Biol. Med.* **111:** 416, 1962.

GRIFFIN, C. W., CARSKI, T. R., AND WARNER, G. S. Labeling procedures employing crystalline fluorescein isothiocyanate. *J. Bact.* **82:** 534, 1961.

HELMKAMP, R. W., GOODLAND, R. L., BALE, W. F., SPAR, I. L., AND MUTSCHLER, L. E. High specific activity iodination of gamma globulin with iodine-131 monochloride. *Cancer Res.* **20:** 1495, 1960.

KABAT, E. A. AND MAYER, M. M. *Experimental immunochemistry* (2d ed.). Springfield, Ill.: Charles C Thomas, 1961, p. 905.

RIGGS, J. L., SEIWALD, R. J., BURCKHALTER, J. H., DOWNS, C. M., AND METCALF, T. G. Isothiocyanate compounds as fluorescent labeling agents for immune serum. *Amer. J. Path.* **34:** 1081, 1958.

RINDERKNECHT, H. A new technique for the fluorescent labeling of proteins. *Experientia* **16:** 430, 1960.

WADSWORTH, C. Comparative testing of a new photographic material for rapid registration of immunoprecipitates. *Int. Arch. Allergy* **23:** 103, 1963.

WEISS, A. J. AND BRENT, R. L. The determination of the biologic lifespan of certain anti-cancer drugs. *Proc. Amer. Ass. Cancer Res.* **3:** 371, 1962.

CHAPTER 10

EMBRYOLOGICAL CONSIDERATIONS
IN TERATOLOGY

JAMES G. WILSON

Several non-embryonic factors of importance in teratology may be mentioned before considering the embryonic ones. First, there are the agent and whatever degree of specificity it may exhibit in producing particular types of malformations. This specificity may not be related to a known mechanism of action but, nevertheless, a considerable correlation can be assumed to exist between the types of malformations produced and specific metabolic events in the developmental history of the affected organ.

Second, there is the matter of dosage, which is not as simple as might at first be assumed. All dosage levels of a known teratogen are not teratogenic. Typically there are a lower range which permits normal development and a higher range which kills all of the embryos—and the mother also if extended far enough. Between these there is a narrow *teratogenic zone,* in which dosage is sufficient to interfere with specific developmental events without destroying the whole embryo. Teratogenic dosage, as will be seen later, is inseparably related to the age of the embryo.

Third, in mammals the maternal organism is the physical environment of the embryo, as well as the immediate source of the metabolites needed for maintenance and growth. Consequently, the physiologic state of the mother is of considerable importance to the embryo (1). Furthermore, the efficiency of maternal-embryonic interchange by way of the placenta could conceivably be a factor in teratogenesis, although such has not been actually proven to be the case. Thus, the nature of the agent, the dosage, the maternal organism, and the effectiveness of the maternal-embryonic exchange are extra-embryonic considerations which may affect teratogenesis.

The strictly embryonic considerations are primarily two, namely, genotype and developmental stage. The importance of genotype has been amply discussed elsewhere (2, 3). In simplest terms genotype is important because

Professor and Chairman of the Department of Anatomy, College of Medicine, University of Florida, Gainesville, Florida.

Supported in part by NIH Grant HD-00607-02.

This is a slight modification of a manuscript submitted for regular publication in the Annals of the New York Academy of Sciences, in press.

it determines the inherent susceptibility of an embryo to a given agent at a given time in development. Embryonic age at the time of teratogenic action is also well-known as an important determinant and has likewise been stressed elsewhere (3). It is such an important factor, however, that further emphasis on certain particular aspects is certainly not out of order.

In general it may be stated that the embryonic stage at the time an agent acts determines which tissues are susceptible to teratogenesis. Susceptibility has been shown to vary greatly during the course of gestation (Plate 1). During

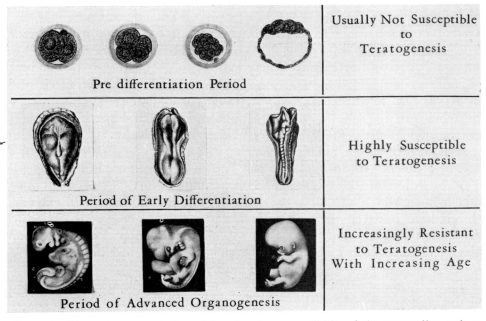

Pre differentiation Period	Usually Not Susceptible to Teratogenesis
Period of Early Differentiation	Highly Susceptible to Teratogenesis
Period of Advanced Organogenesis	Increasingly Resistant to Teratogenesis With Increasing Age

PLATE 1.—Schematic representation of the changing susceptibility of the mammalian embryo. The stages depicted are drawn from, or are retouched photographs of, actual human specimens that have appeared in several papers in the *Contributions to Embryology* of the Carnegie Institution.

cleavage and what would correspond to blastula stages in mammalian embryos, there is typically no teratogenic response to even high doses of agents that at later stages are very effective. This has been repeatedly demonstrated with many agents such as irradiation (4–6), vitamin deficiencies (7, 8), vitamin excess (9), and many others. A few apparent exceptions are known; e.g., Werthemann and Reinigar (10) found ocular defects in a few rats subjected to hypoxia on days 1 to 8 of gestation; Smith (11) noted abnormal development in all members of a single litter of hamsters subjected to hypothermia on day 2 of gestation; Rugh and Grupp (12) described brain abnormality in a very low percentage of mice exposed to ionizing radiations in 1- and 2-cell stages of development. None of these experiments is known to have been confirmed by other workers.

The precise mechanism whereby the great majority of very early embryos resist teratogenic damage is unknown. Whether or not an embryo can be rendered anomalous probably has something to do with whether the component cells have become irrevocably determined to form specific parts of the future organism or whether they retain some degree of the original totipotency of the fertilized egg. As long as all or many cells retain totipotency, destruction or damage to a certain percentage can be tolerated. The embryo is in this case said to *regulate* or undergo internal rearrangements so that normal development may ensue, although very likely with a retardation of developmental schedule proportional to the number of cells lost. Destruction of such relatively undifferentiated cells, then, leaves no teratogenic scar, but there seems to be a critical limit beyond which the damaging of even non-specialized cells cannot be tolerated if the embryo is to live (12, 13).

The onset of teratogenic susceptibility occurs at about the time the germ layers are formed. In mammals this is several days after conception, about 5 days in the hamster and mouse, 8 days in rat, 9 days in rabbit, assumed to be 10 days in monkey, and could be as early as 11 or 12 days in man. In birds susceptibility to teratogenesis starts within a few hours after the beginning of incubation because most of the indifferent stage of development has been passed in the genital tract of the mother bird before the egg is laid. Not only is the onset of susceptibility rather sudden but many teratogenic agents produce their highest incidence of malformations shortly thereafter (4, 5–9, 14). This is at a time when the embryo consists of two or three simple layers of cells arranged as an embryonic disc, with little if any morphological indication of definitive structures. Nevertheless, at these early stages it is possible to demonstrate by transplantation (15) or by chemical means (16) that localized areas of the disc now have acquired specific organ-forming potential, in other words, that induction or chemical differentiation has occurred. Thus, chemically differentiated cells may be subject to teratogenesis several hours or possibly days before their ultimate role in development is indicated by morphological differentiation. For example, renal anomalies can be induced by irradiating rat embryos on the ninth day, but the metanephros does not appear even in rudimentary form until the twelfth day (17), and skeletal abnormalities have been produced by treatment several hours before any structure other than the notochord is recognizable (18, 19).

Most organs have a period of particular susceptibility to teratogenesis, and this very likely coincides with early and critical developmental events in that organ; but as seen above, these events need not be structurally evident.

If more than minimal doses are used, however, most organs can be rendered abnormal at other times as well. In fact, the period during which a malformation may be produced may extend over a few days, although at either end of this time distribution a somewhat larger dosage is required (Fig. 1) (19, 20). To identify the most susceptible period for a given malformation, minimal doses should be used because higher doses may introduce complicating defects or tend to mask the effect by causing prenatal death of some embryos (6, 21). Certain organs such as the palate may be induced to show more than one susceptible period, which calls attention to the fact that there is more than one underlying embryological process involved in palatal closure. These processes may become vulnerable at different times and may respond to different agents.

Susceptibility to teratogenesis decreases as differentiation and organogenesis proceed. Proliferative and morphogenetic activities which character-

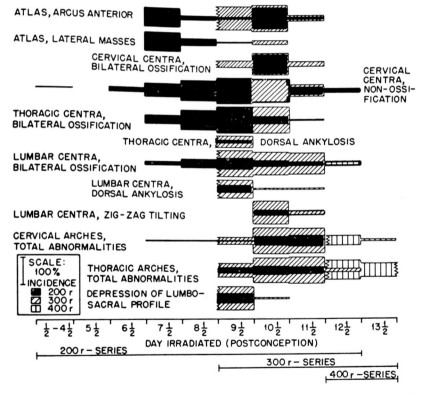

FIG. 1.—Diagram illustrating that maximal susceptibility may be limited to a single day but that the period of susceptibility may extend, albeit at a lower response rate, for a day or two prior, and subsequent to, the time of maximal response. It is also shown that the period of susceptibility may be further extended as well as the incidence of any one malformation increased by increasing dosage. Reproduced with permission from the paper on irradiation effects on mouse development published by L. B. Russell (19).

ize early stages of tissues and organs become less evident, and increasing specialization in structure becomes more apparent as the embryo grows older. As differentiation thus advances, larger doses of the agent are required to produce comparable malformations (6, 22–25), if indeed comparable ones can be produced at all. Not only individual organs but the embryo as a whole becomes progressively more resistant to teratogenesis (4, 8). For example, exposure of rat embryos to 100 r of X-rays on the ninth day of gestation caused many types of malformations in virtually all of the young; the same treatment on the tenth day caused somewhat fewer types of malformations in 75 per cent of the offspring, whereas the same treatment on the eleventh day did not produce any malformations (Table 1) (6).

TABLE 1

INCIDENCE OF MALFORMATIONS PRODUCED BY IRRADIATION
(Expressed as Per Cent)

DAY	Dose (r)			
	25	50	100	200
8....	None	None	None	(No survivors)
9....	Eye, 6	Eye, 72; brain, 9; spinal cord, 3	Eye, 90; brain, 41; spinal cord, 27; heart, 20; face, 14; situs inversus, 13; aortic arch, 10; urinary, 5	Eye, 100; brain, 78; spinal cord, 67; situs inversus, 55; heart, 22; face, 11; aortic arch, 11
10....	....	Eye, 11	Eye, 75; urinary, 11; brain, 3	Eye, 94; feet, 33; brain, 19; urinary, 11; aortic arch, 11
11....	....	....	None	Eye, 100; urinary, 77; brain, 54; spinal cord, 31; aortic arch, 23; ear, 23; tail, 23; heart, 15; jaw, 15; feet, 7

From (6).

The syndrome or pattern of defects associated with an agent may change when the agent is applied at successively later times in gestation, as has been well demonstrated with short-term vitamin deficiencies (8, 26). Such changes reflect the fact that different organs show varying degrees of susceptibility at different times. Hence, a vitamin deficiency limited to days 9 and 10 affects different organs than one limited to days 12 and 13. In fact, it is useful to think of the susceptible period as waxing and waning because abrupt beginnings and endings of susceptibility do not appear to be the rule (Fig. 2). The duration of action of the agent affects the composition of the syndrome as a result of what might be called a cumulative effect; e.g., a vitamin A deficiency which was terminated on the tenth day by therapeutic doses of the vitamin resulted only in heart abnormalities, but allowing the deficiency to continue to the fifteenth day resulted in abnormalities of the eyes, aortic arches, diaphragm, lungs, and several specific genital organs (25). A given organ may show susceptible periods at two or three different times in response to as many agents. For example, Landauer (21) produced the same type of long-bone defects in chicks with boric acid, insulin, and pilocarpine but the peak of susceptibility in the three instances fell on the 96th, 120th, and 144th hours of incubation, respectively.

When organogenesis is completed, the embryo enters the fetal period, which is characterized principally by growth and functional maturation. During this period, teratogenesis in the strict sense does not occur because embryonic processes can no longer be interrupted or diverted. Any agent sufficiently potent to affect the fetus would either retard growth or cause

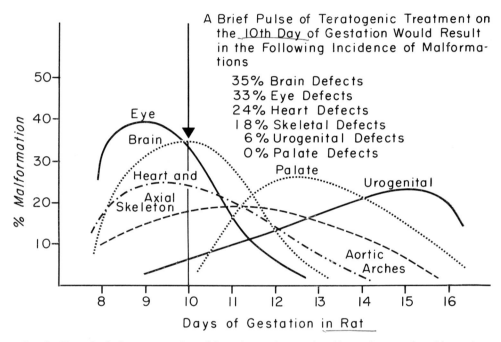

FIG. 2.—Hypothetical representation of how the syndrome of malformations produced by a given agent might be expected to change when treatment is given at different times. The percentage of animals affected as well as the incidence rank of the various types of malformations would be somewhat different from that shown for the tenth day if treatment were given instead on the twelfth or the fourteenth day, for example.

pathology of the types that occur in postnatal animals. Most of the lesions associated with congenital syphilis should, in this light, be regarded as examples of congenital pathology rather than malformations. The developing rat enters the refractory period about the seventeenth day postconception. Human embryogenesis is completed about the end of the eighth week of intrauterine life and, with the exception of external genitalia, developmental defects for the most part cannot be produced after this time.

An illustrative experiment with actinomycin D as the agent in the rat.—

To test and reconfirm some of the generalizations discussed above, the following experiment was undertaken. Rats of the Albino Farms strain, which are Wistar descendants, were selected because they have been used in the author's laboratory for a number of years and found consistently to bear large litters with a low rate of intrauterine mortality and a very low rate of spontaneous malformations. The agent, actinomycin D, an antibiotic with high teratogenic potency, is known to combine with DNA, thereby inhibiting RNA synthesis on the chromosomes, and also to prevent nuclear RNA synthesis in mammalian cells according to Mach and Tatum (27).

The experimental plan was to inject a single dose of actinomycin D intraperitoneally into pregnant rats on one or another day from the sixth through the eleventh of pregnancy. Three dosage levels were used: 0.1, 0.2, and 0.3 mg/kg. As is routinely done in our laboratory, the fetuses were removed on the twentieth day, one day prior to expected term, so that dead and resorbed conceptuses could be counted and abnormal fetuses saved from being eaten by the mother. All living fetuses were examined for gross malformations and preserved for further study. Every third fetus was fixed in 95 per cent alcohol to be cleared by the Schultz-Dawson alizarin red method for subsequent skeletal examination. All others were fixed in Bouin's fluid for 1 week to permit decalcification before making 1-mm free-hand sections with a razor blade. The sections were then examined under a dissecting microscope for internal abnormalities. Although not permitting the minute examination that is possible after histologic serial sections, this combination of methods allowed both evaluation of the skeleton by the clearing technique and of internal organs by the razor blade sections, thus providing adequate information on the types of malformations at the expenditure of only a fraction of the time and energy required for serial sections.

As indicated in Table 2, actinomycin D at 0.1 mg/kg dosage level on the ninth day of gestation is without teratogenic effect, although it did cause an appreciable rise in intrauterine mortality. Since on the ninth day the rat embryo is more sensitive to most chemical agents than at any other time, treatment at other ages with this dose was not attempted. The greater sensitivity at the ninth day was borne out when the dosage level was raised to 0.2 mg/kg. Not only was the total number of animals affected much higher but the number and variety of anomalies per abnormal animal much greater. In fact, virtually every organ system in the body was rendered abnormal occasionally, although the axial skeleton showed abnormalities most frequently.

When the dosage was raised to 0.3 mg/kg, the extremely high lethal effect resulting from the ninth-day treatment precluded evaluation of the teratogenic effect (Table 2). Intrauterine mortality was also high after treatment on the eighth and tenth days, but enough animals survived to make it apparent that this dosage on these days was teratogenically much more effective than the 0.2 mg/kg dosage. The prevailing types of malformations after tenth-day treatment were now ocular and axial skeletal, the eye defects having largely replaced hydrocephalus which was most prevalent after earlier treatment.

Contrary to expectation and to all previous experience with this strain of rats, and a variety of agents, treatment on the seventh day caused a significant incidence of malformations. To add to our consternation, some of the defects were of types rarely seen or never seen before. For example,

TABLE 2

EFFECTS OF ACTINOMYCIN D IN DEVELOPING RATS

Dose (mg/kg)	Day of Treatment	Total Implants	Dead, Resorbed (%)	Survivors Malformed (%)	Most Frequent Defects
0.1	9	212	7.1	1.0	b
0.2	7	122	11.5	1.9	b
0.2	8	120	4.2	16.0	hydrocephalus, axial skeletal
0.2	9	191	32.5	28.1	axial skeletal, hydrocephalus
0.2	10	130	12.3	4.4	b
0.2	11	130	7.7	0.0	....
0.3	6	78	10.3	2.8	b
0.3	7	115	13.0	11.2	hydrocephalus, ectopia cordis
0.3	8	92	84.8	26.6	b
0.3	9	127	99.2	100a	b
0.3	10	109	57.9	65.2	ocular, axial skeletal
0.3	11	124	12.1	0.9	b
None	....	583	3.6	0.9	b

a Only one survivor.
b Too few abnormal animals for frequency evaluation.

ectopia cordis occurred with some regularity but has not been produced by any other agent we have used. Nevertheless, malformations of types regularly produced by actinomycin D were observed, suggesting that aside from the unusual age at which it was obtained this teratogenic response was of a fairly standard type.

A striking parallel was noted between the rates of malformations and of prenatal mortality (Fig. 3). Although not often observed to this degree, some correlation between the teratogenic and the lethal effects has been observed with most agents (3). In some experiments these two effects have appeared to vary independently, whereas in others (as in the present instance) they have been so closely associated as to suggest that teratogenicity and lethality are only different degrees of the same reaction of the embryo to injury.

Investigations are now in progress to learn more about the unusual susceptibility of the early embryo to this agent. The first step will be to determine how far into the pre-differentiation period actinomycin D is capable of causing malformations. If treatment on the seventh day is found to be the earliest at which malformations can be produced, one could postulate that for some reason this agent did not reach the embryo, or at least did not influence the embryo, until the eighth day, thereby preserving the generally accepted principle that rats are not subject to teratogenesis prior to the eighth day. If, however, this agent is found to affect pre-differentiation

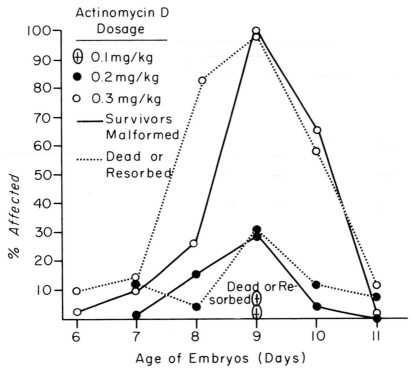

Fig. 3.—Graphic representation of the close parallel between the teratogenic and lethal effects of actinomycin D.

embryos at other or all ages, one would be forced to assume, until proven otherwise, that actinomycin D has a unique mechanism of action which sets it apart from all other commonly used teratogens. This possibility seems not unlikely when it is recalled that actinomycin D has been determined with some certainty to act, not on specific aspects of morphological or even chemical differentiation, but at the site of earliest expression of genetic information. It blocks the synthesis of DNA-dependent RNA on the chromosome. Thus it would prevent or interfere with the initial step of the many chemical and morphogenetic steps required for the development of any tissue or organ. The synthesis of RNA on the chromosomes is known to precede all other demonstrable chemical or structural differentiation (28).

This simple experiment with actinomycin D has re-emphasized several observations and generalizations that were previously known, but it has also introduced a new possibility into experimental teratology. As for the confirmation of earlier generalizations, it has shown (1) that a relatively narrow range of dosage lies between complete normality of fetuses, on the one hand, and complete lethality, on the other, and that this range is the teratogenic zone wherein the rates of mortality and teratogenicity tend to follow parallel courses as dosage and time of treatment vary;

(2) that the embryo is most susceptible to teratogenesis during the early stages of morphological differentiation, which in the rat falls on the eighth and ninth days, after which sensitivity declines very rapidly until the eleventh day when there is no response to a dose of antinomycin D (0.3 mg/kg) sufficient on the ninth day to cause 100 per cent death and/or abnormality;

(3) that during the short span between the eighth and tenth days the composition of the syndrome of malformations changes appreciably, e.g., from hydrocephalus as the prevailing type on the eighth day, to predominantly axial skeletal defects on the ninth day, to ocular defects as the commonest type on the tenth day, with the incidence ranks of the accompanying less common types also changing.

The unusual finding was that treatment prior to the eighth day could cause significant numbers of malformations. Previous workers have almost without exception been unable to produce malformations on the seventh day, although high mortality of embryos has been regularly reported when adequate dosage was used. Actinomycin D not only produced malformations on the seventh day, but it produced some types not seen after treatment at later stages. It is postulated that this agent acts not only by interfering with specific aspects of tissue or cellular differentiation, but even earlier by interfering with chromosomal differentiation, i.e., the synthesis of specific RNA on the chromosome, hence its ability to act earlier than other agents.

REFERENCES

1. WILSON, J. G. Influence on the offspring of altered physiologic states during pregnancy in the rat. *Ann. N.Y. Acad. Sci.* **57**: 517, 1954.
2. FRASER, F. C. Genetics and congenital malformations. In: *Progress in medical genetics.* New York: Grune and Stratton, 1961, pp. 38–80.
3. WILSON, J. G. General principles of experimental teratology. *Congenital malformations.* Proceedings of the First International Conference. Philadelphia: Lippincott, 1961.
4. HICKS, S. F. Developmental malformations produced by radiation. A timetable of their development. *Am. J. Roentgenol.* **69**: 272, 1953.
5. RUSSELL, L. B. X-ray induced developmental abnormalities in the mouse and their use in the analysis of embryological patterns. I. External and gross visceral changes. *J. Exp. Zool.* **114**: 545, 1950.

6. WILSON, J. G. Differentiation and the reaction of rat embryos to radiation. *J. Cell. Comp. Physiol.* (Suppl. 1), **43**: 11–38, 1954.

7. CHENG, D. W., CHANG, L. F., AND BAIRNSON, T. A. Gross observations on developing abnormal embryos induced by maternal vitamin E deficiency. *Anat. Rec.* **129**: 167, 1957.

8. NELSON, M. M., WRIGHT, H. V., ASLING, C. W., AND EVANS, H. M. Multiple congenital abnormalities resulting from transitory deficiency of pteroylglutamic acid during gestation in the rat. *J. Nutr.* **56**: 349, 1955.

9. COHLAN, S. Q. Congenital anomalies in the rat produced by excessive intake of vitamin A during pregnancy. *Pediat.* **13**: 556, 1954.

10. WERTHEMANN, A. AND REINIGER, M. Uber augenentwicklungsstorungen bei rattenembryonen durch sauerstoffmangel in der fruhschwangerschaft. *Acta Anat.* **11**: 329, 1950.

11. SMITH, A. U. The effects on foetal development of freezing pregnant hamsters (*Mesocricetus auratus*). *J. Embryol. Exp. Morph.* **5**: 311, 1957.

12. RUGH, R. AND GRUPP, E. Response of the very early mouse embryo to low levels of ionizing radiations. *J. Exp. Zool.* **141**: 571, 1959.

13. WILSON, J. G., BRENT, R. L., AND JORDAN, C. H. Differentiation as a determinant of the reaction of rat embryos to X-irradiation. *Proc. Soc. Exp. Biol. Med.* **82**: 67, 1953.

14. INGALLS, T. H. AND CURLEY, F. J. Principles governing the genesis of congenital malformations induced in mice by anoxia. *N. England J. Med.* **257**: 1121, 1957.

15. RAWLES, M. E. A study in the localization of organ-forming areas in the chick blastoderm of the head-process stage. *J. Exper. Zool.* **72**: 271, 1936.

16. EBERT, J. D., TOLMAN, R. A., MUN, A. M., AND ALBRIGHT, J. F. The molecular basis of the first heart beat. *Ann. N.Y. Acad. Sci.* **60**: 968, 1955.

17. WILSON, J. G., JORDAN, H. C., AND BRENT, R. L. Effects of irradiation on embryonic development. II. X-rays on the ninth day of gestation in the rat. *Am. J. Anat.* **92**: 153, 1953.

18. MURPHY, L. M. A comparison of the teratogenic effects of five polyfunctional alkylating agents on the rat fetus. *Pediat.* **23**: 231, 1959.

19. RUSSELL, L. B. X-ray induced developmental abnormalities in the mouse and their use in the analysis of embryological patterns. II. Abnormalities of the vertebral column and thorax. *J. Exp. Zool.* **131**: 329, 1956.

20. WARKANY, J. AND SCHRAFFENBERGER, E. Congenital malformations induced in rats by roentgen rays. *Am. J. Roentgenol. Rad. Ther.* **57**: 455, 1947.

21. LANDAUER, W. On the chemical production of developmental abnormalities and of phenocopies in chicken embryos. *J. Cell. Comp. Physiol.* **43**: 261, 1954.

22. DAGG, C. P. AND KARNOFSKY, D. A. Teratogenic effects of azaserine on the chick embryo. *J. Exp. Zool.* **130**: 555, 1955.

23. FRASER, F. C. AND FAINSTAT, T. D. Production of congenital defects in the offspring of pregnant mice treated with cortisone. *Pediat.* **8**: 527, 1951.

24. KALTER, H. The inheritance of susceptibilty to the teratogenic action of cortisone in mice. *Genetics* **39**: 185, 1954.

25. WILSON, J. G., ROTH, C. B., AND WARKANY, J. An analysis of the syndrome of malformations induced by maternal vitamin A deficiency. Effects of restoration of vitamin A at various times during gestation. *Am. J. Anat.* **92**: 189, 1953.

26. NELSON, M. M., BAIRD, C. D. C., AND WRIGHT, H. V. Multiple congenital abnormalities in the rat resulting from riboflavin deficiency induced by the antimetabolite galactoflavin. *J. Nutr.* **58**: 125, 1956.

27. MACH, B. AND TATUM, E. L. Ribonucleic acid synthesis in protoplasts of Escherichia coli; inhibition by actinomycin D. *Science* **139**: 1051, 1963.

28. COWDEN, R. R. AND LEHMAN, H. E. A cytochemical study of differentiation in early echinoid development. *Growth* **27**: 185, 1963.

METHODS FOR ADMINISTERING AGENTS AND DETECTING MALFORMATIONS IN EXPERIMENTAL ANIMALS

METHODS OF ADMINISTERING CHEMICAL AGENTS

The advantages and disadvantages and some suggestions regarding several methods for administering chemical teratogenic agents to laboratory mammals are summarized below.

Incorporation in the diet.—This is useful with agents that are to be given in large amounts or are difficult to dissolve in vehicles that would be tolerated in other treatment routes. It has two disadvantages: (*1*) the dosage is difficult to measure even when the percentage in the diet is carefully controlled, unless considerable effort is made to determine the animal's actual food consumption, and (*2*) the composition of the agent may be altered by action of digestive enzymes.

Gavage.—A stomach tube is simple to administer to most laboratory animals if gag and swallowing reflexes are suppressed with mild anesthesia. This method is particularly useful with oily solutions or suspensions, particularly if they need to be given in large quantity or frequent dosage. Virtually insoluble compounds such as thalidomide can be effectively given as a suspension in vegetable or mineral oil by this means. Regurgitation is usually not a problem when mild anesthesia is used and care is taken to make sure that the tip of the tube is in the stomach when the dose is delivered. Urinary catheters can be adapted to this purpose but a practically foolproof stomach tube for small animals can be made by soldering a small lead bead to the tip of a No. 16 spinal-tap needle from which the sharp point has been filed. The latter type of stomach tube has the advantage that it can be inserted with a loaded syringe attached.

Unction.—This method has not been widely employed in teratological studies, but it has been used to apply sex hormones to the pouch-borne fetuses of the opossum. Oily solutions of hormones were rubbed directly on the skin of the very immature young shortly after they were transferred from the uterus to the pouch.

Injections.—These may be made at several locations with varying degrees of facility and effectiveness.

SUBCUTANEOUS INJECTIONS are easy to do, often without anesthetizing the animal, and have the advantage that doses of moderately large volume can be given at multiple sites or by redirecting the needle from a single skin puncture. Agents and vehicles of surprisingly varied and unphysiologic nature are tolerated by this route. Absorption is relatively slow, particularly when an oily vehicle is used.

INTRAMUSCULAR INJECTIONS offer a somewhat faster absorption time than no ?
subcutaneous treatment, but the volume of material that can be delivered
at one site is somewhat less, which may account for the fact that this method
has been relatively little used in the smaller laboratory animals.

INTRAPERITONEAL INJECTION insures very rapid absorption of most agents
and vehicles and quite large injection volumes can be used if reasonable
precautions regarding isotonicity, pH, and temperature are taken. Although
not essential, light anesthesia minimizes the likelihood of puncturing abdominal viscera. Peritonitis is a possible complication of irritating substances
are injected, a viscus is punctured, or needles, syringes, and solutions are
not sterile.

INTRAVENOUS INJECTION brings the agent to the embryo almost instantane-
ously and in relatively high concentration, hence it is of value when an
abrupt onset of action is desirable. This is not a suitable method when the
material to be injected is likely to be thrombogenic or when the injection
mass is large. To enter a superficial vein with consistency and without occasional loss of part of the dosage requires some degree of skill, particularly
in small animals such as rats and mice. In these two species the most readily
accessible vessels are the caudal veins of the tail. Because of the thinner
skin and consequent greater visibility of the caudal veins, this type of treatment is usually easier in mice than rats. The tail veins of rats can be made
more visible by suspending the tail in a beaker of warm water for a minute
prior to injection or by applying an irritant such as xylene to cause vasodili-
tation. A ¾-inch 25-gauge needle is most suitable.

METHODS OF EXAMINING RAT FETUSES FOR MALFORMATIONS

External examination.—Newly delivered maternal rats and mice tend to
eat their defective offspring, particularly if the defective individuals have
a surface lesion from which blood or body fluids escape or if the young are
dead or in a moribund state at birth. To evaluate accurately a teratological
result, therefore, the pregnant animals must be sacrificed before full term
and since term varies over a range of several hours, it is good practice routinely to interrupt pregnancy 12 to 24 hours before expected delivery. Any
method of killing the mother is satisfactory that does not involve physical
trauma to the fetuses. An overdose of ether is one of the cleanest and most
convenient ways.

The abdominal wall of the mother should be opened and the full extent
of both uterine horns exposed promptly. Before opening either horn, however, a careful count should be made of both the fetal swellings and of the
"metrial glands." Metrial glands are highly vascularized yellowish nodules
which are found along the mesometrial margin of the uterine horns where

they mark any original implantation site, whether the embryo or fetus associated with that site survives or not. Accordingly there is a metrial gland at the base of each placental attachment, but there may be metrial glands which are no longer associated with placentae and which represent sites at which embryos or fetuses have undergone prior death or resorption. A recognizable remnant of the placenta may or may not remain, presumably depending on the time elapsed since resorption occurred. Animals which have multiple implantations, such as rats and mice, resorb (re-absorb) dead conceptuses *in situ* rather than abort them as in the case in primates and ungulates. In any event, the metrial glands unoccupied by living or recently dead fetuses represent the number of prior resorptions. It has been repeatedly shown that most agents produce lethal effects that parallel or at least vary in the same direction as the teratogenic effects, and the number of resorbed or dead conceptuses may provide an interesting correlate to the extent of teratogenicity. After the numbers of resorbed and intact offspring are counted and recorded, the uterus should be opened with as little delay as possible to determine which fetuses are alive and which dead but intact. Spontaneous movement and a more ruddy color will usually distinguish living from dead fetuses, but in case of doubt gentle pressure applied with a forceps or other instrument about the head or neck will elicit reflex movement from otherwise inactive individuals.

A consistent system of numbering and indicating the relative positions of fetuses should be adopted. If the mothers have been numbered, it may avoid confusion to assign letters to the fetuses and resorption sites. A convenient system is to begin at the left ovary and assign letters *A, B, C,* etc., to each implantation site in order down the left horn, across the cervix and up the right horn to the right ovary. Serial listing of the fetuses on a protocol sheet (Fig. 1) will automatically record the relative positions of all living, dead, and resorbed fetuses and sites within the uterus, and within each horn if a line or arrow is drawn to indicate the position of the cervix.

Living young should be removed consistently in whatever order is adopted. A small hemostat or artery clamp should be placed on the umbilical cord flush with the abdominal wall and the cord broken or cut distal to the clamp which can then serve as a handle. Lift the fetus onto blotting paper to remove excess amniotic fluid and blood before finally placing it on the balance for weighing. The clamp can now be removed with little further danger of bleeding from the cord. Weighing near-term rat or mouse fetuses to more than 0.1 g is probably a waste of time. Abnormal fetuses are not necessarily smaller than their normal littermates, but fetal weights should be taken because they give information relative to developmental state of the entire litter, general health, and nutritional state of the mother.

External examination of the fetus should proceed in an orderly manner

Date Received_____

Serial No._____

Ear_____ Toe_____

Type Expt._____

gestational age(s)_____

Day 1 of Treatment_____ Wt_____

Date

Treatment Details_____

_____

_____

Laparatomy Date_____

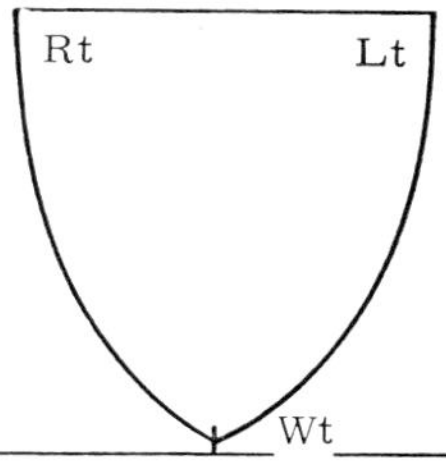

Rt   Lt

Autopsy, date_____ Wt_____

Remarks_____

Date	Mixture	Leukocytes	Epithelial	Cornified	Bred	Male	Sperm Present?	Wt(g)

Fetus	Sex	Weight	Gross Appearance	Fix.	Subsequent Study
1					
2					
3					
4					
5					
6					
7					
8					
9					
10					
11					
12					
13					
14					
15					

FIG. 1.—Protocol sheet used for recording breeding, treatment, autopsy, and fetal information.

from head to tail and, although not necessary, may be done with greater ease under a long focal-length lens of about 2–3 magnification. The contour of the cranium should first be noted in profile and face-on view, and with a little experience one quickly learns to recognize the dome-shaped configuration typical of hydrocephaly. Retarded fetuses, however, may be misleading in that the cranium rises proportionately higher than in more mature animals. Under proper lighting the eye bulges are quite prominent in most near-term laboratory animals. Not only should the bilateral presence of eye bulges be noted, but it should be determined whether they are of equal size and position. Anophthalmia, microphthalmia, and some forms of frank ocular malformation may be recognized on external examination. Easily identified are such combined cranial and brain defects as exencephaly, in which the cranium and skin are absent and the brain is exposed, usually everted, on the surface of the head; meningoencephalocele, in which the skin is intact but the brain or some part of it together with meninges protrudes through a cranial defect to cause an irregular mass beneath the skin; and simple meningocele in which the skin is intact and translucent and is elevated by a fluid filled vesicle of meninges which protrudes through a midline defect in the cranium. The face or muzzle should be examined face-on for gross distortions and for the less conspicuous notches or furrows that indicate some degree of cleft lip, cleft mandible, or oblique facial cleft. Cleft palate, if extensive, can be seen by inserting a small forceps in the mouth and forcing the jaws apart, but incomplete forms may be visible only after sectioning. Abnormalities of the external ear are manifested by variations in size, shape, and position. The latter may or may not be associated with micrognathia (short mandible) and more severe mandibular defects.

Abnormalities of the head, particularly those affecting the eyes and brain, are more likely to be encountered in animals treated early in the period of teratological susceptibility (7–8 days in mice, 8–9 days in rats).

Abnormalities of the extremities are usually indicated externally by unusual size or position of the limbs or by the number and disposition of digits. Toes should be routinely counted and the depth of the digital furrows noted so as not to overlook such obvious defects as syndactyly, polydactyly, and adactyly. Club foot is evidenced by distortion in the distal parts of an extremity, but care must be taken to rule out spurious situations such as the contorted posture of a limb that can result from leaving the fetus too long in an unopened uterus after killing the mother. Micromelia and phocomelia are easily identified by comparing proportions and dimensions of limbs with those of control fetuses. Malformations of the extremities occur more often but not exclusively when treatment is given later in the teratogenic period (12–14 days in rats and mice).

Abnormalities of the trunk region are usually quite conspicuous, e.g., umbilical hernia in which one or a few loops of intestine protrude through the umbilical opening; and gastroschisis in which some if not most abdominal viscera are outside the defective abdominal wall. Defects along the dorsal midline are also easily identified, a bubble-like bulge if a myelomeningocele or a raw, usually bloody, depression if a spina bifida. If several vertebrae above the caudal level are missing, the trunk may appear shorter and chubbier than usual. Scoliosis and other abnormal vertebral curvatures cannot be detected with certainty on external examination because near-term fetuses may assume unusual postures for prolonged periods, particularly if they are chilled. The caudal vertebrae are relatively often subject to variation, resulting in short, curly, angular, or absent tail.

Sex should always be determined, not that external genitalia are often abnormal but because on occasion there are discrepancies between internal and external genitalia. It has also been claimed that certain teratogenic agents (X-rays) alter the expected sex ratio. In any event, sex is easily determined by observing the distance between the anus and the genital tubercle; it is about twice as great in the male as in the female, e.g., 2 mm in newborn male rats, 1 mm in newborn female rats. Anal atresia has been noted in newborn rats that also have extensive vertebral defects after treatment with trypan blue.

Preservation.—After external examination all animals should be preserved or fixed for further study and the type of preservation depends on the type of further study to be undertaken. If histological sections are to be made, 10 per cent formalin or Bouin's fluid are acceptable although Bouin's fluid has two distinct advantages: (*1*) it causes less shrinkage; and (*2*) it is sufficiently acid that it decalcifies bony structures within a week or two, thus greatly facilitating the sectioning process. Zenker's fluid is also usable, but its advantage in enhancing differential staining is mostly lost on fetal material which already has strong stain affinities, and it has the disadvantage of causing hardening and requiring extensive washing after fixation.

Undoubtedly the most exhaustive search for congenital malformations is made on histologic serial sections. A question can be raised, however, as to whether the tedious and time-consuming procedures required to serially section and study one near-term fetus are justified by the additional information thereby made available. This question must be answered in terms of the objectives of the individual investigator. In most situations in which the primary needs are to know whether a given agent has caused abnormal development, the approximate numbers of fetuses affected, and in general terms, what organs are affected, a less laborious method of study may be adequate.

Two relatively easy methods that yield considerable information about the normality of internal structure are available. One is the technique of

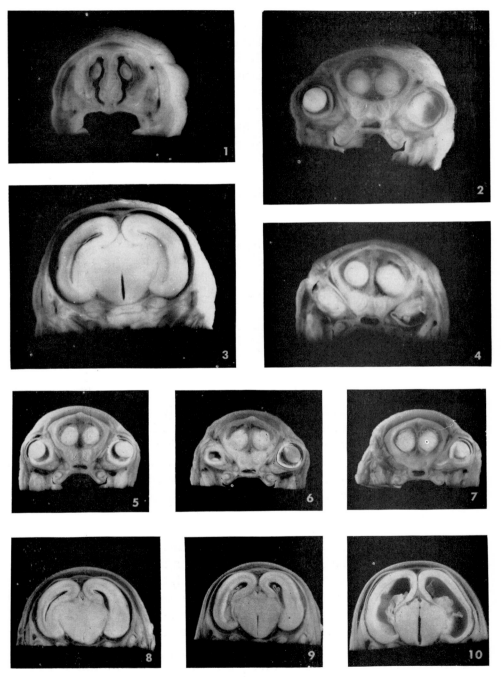

PLATE 1

making free-hand razor blade sections which are studied, and dismembered if necessary, under the dissecting microscope. The other is one of the many techniques by which skeletal structure is made visible by clearing soft tissues and staining with alizarin red. Obviously both methods cannot be used on the same fetus or newborn, but by random or judicious selection of specimens to be sectioned or cleared, it is usually possible to combine these methods to provide a reliable picture of the over-all effects of a given teratogen. In the author's laboratory every third fetus is routinely fixed in 95 per cent alcohol preparatory to use of the clearing procedure and the remaining fetuses are preserved in Bouin's fluid for subsequent free-hand sectioning. This routine is modified only when a fetus that would otherwise be sectioned is suspected of having an abnormality that could best be verified by studying the skeleton of a cleared specimen, e.g., missing vertebrae or defective radius, or an animal scheduled to be cleared is thought to have a soft tissue defect such as hydrocephalus which can best be verified by sectioning. Admittedly this alternation of method would permit an unsuspected and infrequent anomaly of soft tissue to be missed one time in three and an unsuspected and infrequent skeletal anomaly to be missed two times out of three. Nevertheless, it has the great advantage of providing information on a whole litter with relatively little effort or expense and after the lapse of no more than 2 weeks (most of this time is consumed by clearing or decalcification). To prepare histologic serial sections of a single fetus requires at least

---

SECTIONS THROUGH HEADS OF NORMAL AND ABNORMAL ANIMALS

(All sections oriented so that rostral surface is toward viewer, dorsal is toward top of page, and the animal's right is at the viewer's left.)

Section 1.—Palate and nasal cavities. Note horizontal continuity of palate and vertical alignment of nasal septum.

Section 2.—Through eyeballs and olfactory bulbs. The lens has been removed from the animal's left eye to expose the inner surface of the retina. Note size and uniform texture of lens, even curvature of retina, and crescentic shape of vitreous chamber.

Section 3.—Through brain at greatest transverse diameter of head. Note slitlike lateral and third ventricles. At other levels of section, lateral ventricles will vary in size and shape, but with experience one quickly learns the normal ranges.

Section 4.—Compare with Section 2. Both eyes are severely malformed, e.g., lenses are absent, retinae are folded, and vitreous chambers are irregular or absent.

Section 5.—Normal eyes for comparison with Sections 6 and 7.

Section 6.—Right eye is abnormal; e.g., the lens is missing, the eyeball is small and irregular, and a fold in the retina projects into the vitreous chamber. The left eye is normal.

Section 7.—Anophthalmia on right side, moderate microphthalmia on left. To make sure the plane of the cut did not miss the right eye, the section should be dissected with iridectomy scissors and fine-pointed forceps.

Section 8.—Normal brain ventricles that represent the upper limit of normal for size of lateral ventricles. Compare with Sections 9 and 10.

Section 9.—Internal hydrocephalus of moderate degree affecting only lateral ventricles.

Section 10.—Full-blown internal hydrocephalus affecting lateral ventricles to extreme degree but third ventricle only to moderate degree.

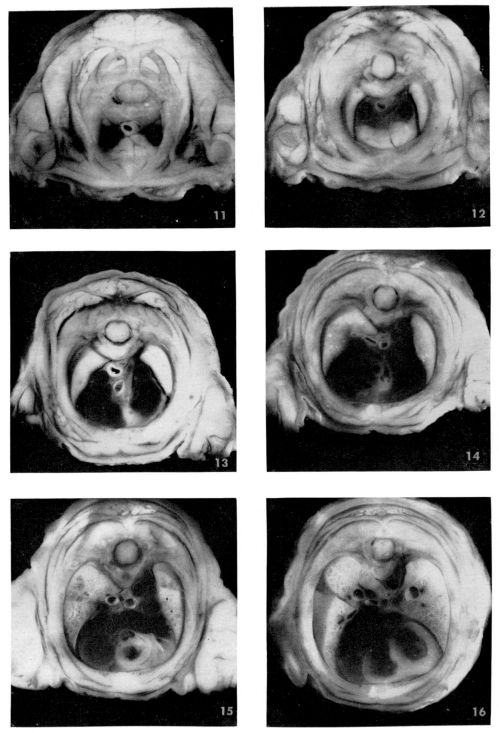

PLATE 2

3 weeks of a technician's time and a considerable outlay for slides, cover glasses, and reagents. The author has not attempted a comparative time study, but after having processed numerous embryos by both histologic serial sectioning, on the one hand, and combined razor-blade section and clearing procedures on the other, he would estimate that the technician-hours required per litter would be in the ratio of about 20 hours for the former for 1 hour of the latter.

Free-hand razor blade sections.—As soon after removal from the uterus as is practical, fetuses are placed in about 25 cc of Bouin's fluid (Allen's modification preferred) per animal and allowed to remain for at least 1 week, which is usually sufficient time for the proper degrees of hardening and decalcification. If on sectioning the fetuses are found to be "rubbery" or to contain bone, they should be returned to the fixative for a few additional days. Before sectioning, each animal should again be given a check for external malformations. The legs and tail should then be clipped off with scissors where they join the trunk. Holding the neck firmly with large forceps or between the fingers, place the fetus in a supine position on a flat block of paraffin. Insert a sharp, single-edge razor blade between the jaws and with firm, even, back-and-forth strokes cut off the upper part of the head in a plane just above the ears. Examine the under side of the palate by removing the body of the tongue which may have remained with the upper part of the head. Place the cut surface of the head down on the paraffin block and begin making 1-mm transverse slices through the head beginning just in front of the eye and proceeding slice-by-slice backward to the region of the ear. This

---

SECTIONS THROUGH NORMAL ANIMALS

(All sections oriented so that cephalic surface is toward viewer, dorsal is toward top of page, and the animal's right is at the viewer's left.)

Section 11.—Lower neck, showing trachea, esophagus, and thymus.

Section 12.—Upper thorax, showing arch of aorta coursing diagonally to the left of the trachea and esophagus. The apices of the lungs appear on either side, and the thymus is still present in the ventral part of the chest cavity.

Section 13.—Plane of sectioning passes caudal to arch of aorta and cephalic to entry of ductus arteriosus into descending aorta. The exterior of the ductus is seen coursing to the left of the ascending aorta. Lungs are conspicuous on either side. Dark areas of the section are occupied by the atria of the heart, which are filled with clotted blood.

Section 14.—The pulmonary trunk with its valve is seen to the left of the ascending aorta. It divides into right and left pulmonary arteries just ventral to the bronchi, which have arisen from the trachea since the last section.

Section 15.—The aortic valve and the pulmonary conus (cephalic part of right ventricle) are sectioned. Two bronchi lie ventral to the esophagus. The right lung in this and subsequent sections is conspicuously larger than the left. Dark areas of the section are occupied mostly by the atria and major veins, both of which are so variable in size and position as to be difficult to study.

Section 16.—Right and left cardiac ventricles and the interventricular septum are shown. The interatrial septum is usually not readily apparent, and since it is open until after birth, no effort is made to identify it. Two of the four lobes of right lung are seen. The left lung normally has only one lobe.

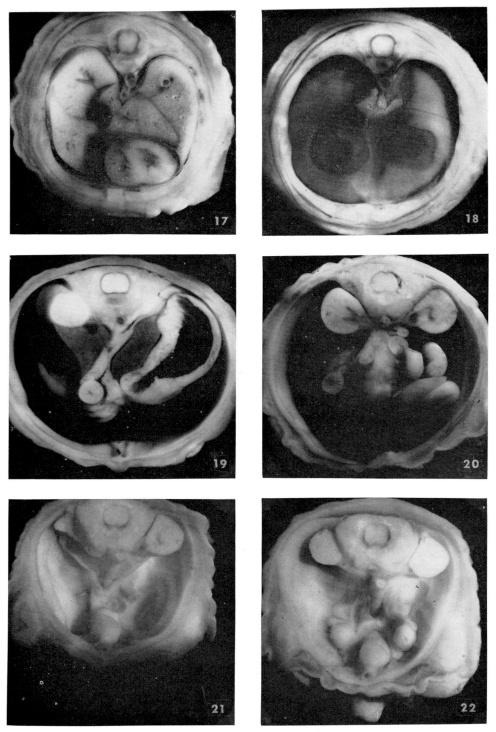

PLATE 3

should yield 5 or 6 1-mm slices, each of which should be placed in individual compartments of a white porcelain spot-test dish and covered with 70 per cent alcohol. The dish containing the sections can be placed on the stage of a dissecting microscope and the sections examined at 5–7.5 × total magnification in the order in which they were made. After a little experience one quickly acquires enough skill to make uniform sections through almost any desired region. Such things as the variable size and shape of the brain ventricles may at first be confusing, but a set of slices through the normal brain can be kept handy for reference in case of questionable ventricular enlargement (hydrocephalus). If certain sections, e.g., through the eyes, do not show internal structure well, a point in doubt may often be settled by literally dissecting the section under the microscope, using iridectomy scissors and fine-pointed forceps. In summary, sections of the head should reveal abnormalities of the palate, nasal cavities, eyes, and abnormalities of the brain, such as internal hydrocephalus.

After the study of the head is complete, the trunk should again be placed supine on the paraffin block and transverse 1-mm-thick sections made begin-

---

SECTIONS THROUGH NORMAL ANIMALS

(All sections oriented so that cephalic surface is toward viewer, dorsal is toward top of page, and the animal's right is at the viewer's left.)

Section 17.—Cardiac ventricles have been cut near the apex of the heart. The lobe of lung dorsal to the heart belongs to the right lung as do the two lobes occupying the right side of the chest. Any of several degrees of situs inversus of lungs would be visible at this level. Esophagus and descending aorta lie near the midline ventral to the vertebral column.

Section 18.—The highest part of the dome of the diaphragm has been cut on either side, and the liver is seen through the openings thus created. The remainder of the diaphragm is seen on its thoracic surface after remnants of lung have been removed.

Section 19.—Somewhat caudal to the previous section bits of liver may still be present, but a conspicuous structure on the left side is the stomach. Ventrally and near the midline are duodenum and pancreas. The cephalic part of the right kidney has been sectioned.

Section 20.—Both kidneys have been cut through the renal pelvis. The size and configuration of the pelves (both normal here) are important clues to the recognition of hypoplastic kidneys, in which the pelvis will be collapsed, or lower urinary tract obstruction, in which the pelvis will be distended. The side-to-side as well as the cephalocaudal positioning of the kidneys should be noted since ectopia in both directions may occur. Little can be gained by studying the sectioned loops of intestine except to note the presence of bile-stained meconium, the absence of which would suggest atresia of the intestine caudal to entry of the bile duct.

Section 21.—When intestines are removed from the caudal end of the trunk, a good view of pelvic organs is obtained. The bladder attached to the ventral body wall is easily recognized. In this female specimen, the two uterine horns are seen to emerge from the pelvic floor dorsal to the bladder and course along the dorsal body wall toward the kidney. On the right, the kidney remnant has been removed to reveal the right ovary. On the left, the ovary is obscured by the caudal part of the left kidney. The rectum is found near the midline, and, although not visible here, the ureter can usually be seen running caudally on either side of the rectum.

Section 22.—The male pelvis is readily identified from the female by the testes on either side of the bladder in the pelvic floor. If testes were at a more cephalic position on the posterior body wall, i.e., nearer the kidneys, it would signify cryptorchidism. Ureters can be seen on the medial aspect of the kidneys in this section.

Detecting Malformations

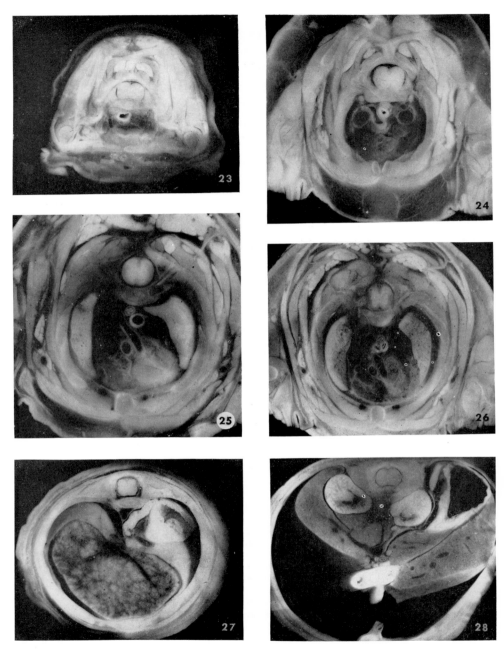

PLATE 4

ning in the region of the shoulder joint. (Sections through the lower jaw and neck are usually not made because these regions contain little of teratological interest, at least that can be studied by this method.)

Sectioning should continue caudally until it is apparent that the kidney region has been passed. Thoracic and abdominal sections are usually more nearly parallel and of uniform thickness if the forceps with which the trunk is held are placed just caudal to the plane of the sectioning as each section is made. Too much pressure on the forceps will cause displacement of viscera and consequent distortion of the sections. It is helpful to make the sections through the heart region a bit thinner than 1 mm in order to reveal as many of the complex septal and valvular structures as possible. Those not revealed directly by the sectioning, however, can often be seen to good advantage by digging the clotted blood out of the sections with a sharp forceps or probe. Actual dissection with iridectomy scissors may also be helpful. Experience will soon indicate that sections through the upper abdomen contain little except liver and, therefore, may be cut somewhat thicker than 1 mm. Thin slices are again desirable in the kidney region to insure at least one cut through the renal pelves. Beyond the kidneys no sections are needed because the genitourinary organs in the pelvic floor and along the posterior body wall are adequately seen by removing the loosely attached intestines and looking directly into the pelvic cavity with the dissecting microscope.

Many malformations can be identified in sections through the trunk. Thoracic sections have revealed right-sided arch of aorta, right-sided ductus arteriosus, dextrocardia, cor biloculare, interventricular septal defects, transpositions of the great arteries, truncus communis, tracheoesophageal fistulae, situs inversus of the lungs, and herniated abdominal viscera resulting from diaphragmatic defects. Abdominal sections have shown diaphragmatic herniae, situs inversus of abdominal viscera, ectopic kidneys, fused kidneys,

---

SECTIONS SHOWING VARIOUS ABNORMALITIES

(All sections oriented so that cephalic surface is toward viewer, dorsal is toward top of page, and the animal's right is at the viewer's left.)

Section 23.—Tracheoesophageal fistula. In this section the esophagus is seen entering the left dorsal aspect of the tracheal wall. At more caudal levels, the esophagus is missing altogether. Compare with Section 11.

Section 24.—Right-sided arch of aorta in which the most cephalic part of the arch is seen passing to the right rather than the left side of the trachea and esophagus. Compare with Section 12.

Section 25.—Right-sided ductus arteriosus and descending aorta. The ductus is seen passing to the right side of the trachea and esophagus and emptying into the descending aorta which descends on the right rather than the left side of the esophagus. Compare with Sections 13 and 14.

Section 26.—Same abnormality as in Section 25, but this section is at a slightly more caudal level.

Section 27.—Diaphragmatic hernia in which the stomach and part of the pancreas protrude into the thorax through a defect in the left side of the diaphragm. Compare with Section 18.

Section 28.—Ectopia of left kidney. The right kidney is normally situated, but the left is displaced ventrally and toward the midline. Compare with Section 20.

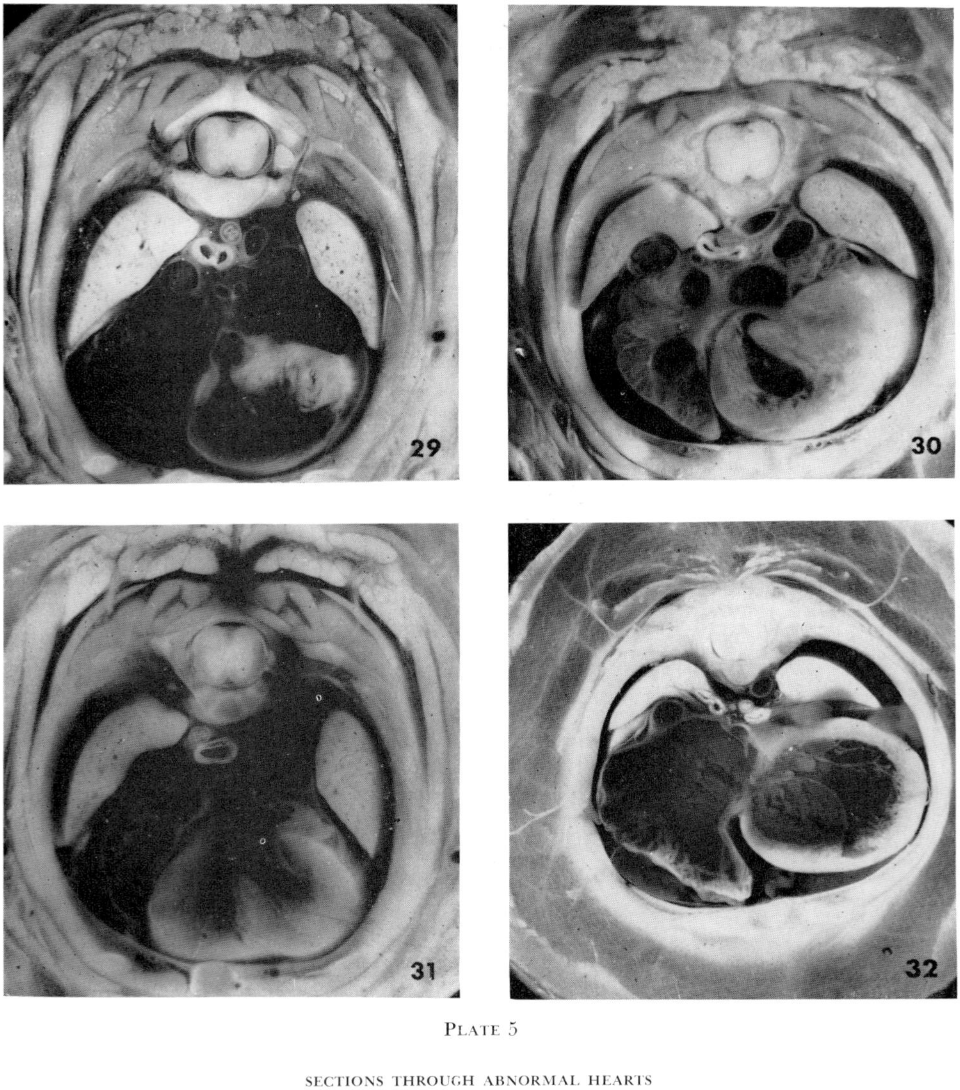

PLATE 5

SECTIONS THROUGH ABNORMAL HEARTS

(All sections oriented so that cephalic surface is toward viewer, dorsal is toward top of page, and the animal's right is at the viewer's left.)

Section 29.—Fistulous connection between right ventricle and ascending aorta at the level of the aortic valve. Compare with Section 15.

Section 30.—High interventricular septal defect which permits communication between the pulmonary conus of the right ventricle and the aortic vestibule of the left ventricle. Compare with Sections 15 and 16.

Section 31.—Large interventricular septal defect. Compare with Section 16.

Section 32.—Cor biloculare, in which the cardiac septa failed to develop and the original atrial and ventricular chambers persist, with no indication of partitioning into the usual four chambers, hypoplastic kidneys, hydronephrosis, and renal agenesis. By looking into the pelvic cavity without sectioning, one can see hydroureters, cryptorchid testes, and various degrees of hermaphroditism.

A series of typical sections is pictured here. With these pictures and after having sectioned several normal fetuses, an intelligent technician or student assistant should be able to detect abnormality, although he should be instructed always to consult with the investigator in case of doubt.

Clearing techniques.—Some of the various modifications of the clearing-alizarin red methods for skeletal study have been described elsewhere in this volume. This author has had excellent results using KOH as the clearing agent after initial fixation in 95 per cent alcohol. The speed of clearing can be varied over a rather wide range depending upon the percentage of KOH used; the usual procedure is 25 cc of 1–2 per cent aqueous solution per 20-day rat fetus.

LIST OF MEMBERS OF THE COMMISSION
ON DRUG SAFETY

L. T. COGGESHALL, M.D., Vice-President of the University, The University of Chicago, CHAIRMAN

PAUL R. CANNON, M.D., Editor, *Archives of Pathology*, American Medical Association

THOMAS FRANCIS, JR., M.D., Chairman of the Department of Epidemiology, School of Public Health, The University of Michigan

PHILIP S. HENCH, M.D., Emeritus Professor of Medicine, The University of Minnesota; Nobel Laureate

HUGH H. HUSSEY, JR., M.D., Director, Division of Scientific Activities, American Medical Association

CHESTER S. KEEFER, M.D., Wade Professor of Medicine, Boston University

THEODORE G. KLUMPP, M.D., President and Director, Winthrop Laboratories

JOHN T. LITCHFIELD, JR., M.D., Director of Research, Lederle Laboratories

MAURICE R. NANCE, M.D., Medical Director, Smith Kline and French Laboratories

LEONARD A. SCHEELE, M.D., Senior Vice-President, Warner-Lambert Pharmaceutical Company

LEON H. SCHMIDT, PH.D., Director, National Primate Center, The University of California

AUSTIN SMITH, M.D., President, Pharmaceutical Manufacturers Association

THOMAS B. TURNER, M.D., Dean of the School of Medicine, The Johns Hopkins University

JOSEF WARKANY, M.D., Professor of Research Pediatrics, Children's Hospital Research Foundation, The University of Cincinnati

———————————

DUKE C. TREXLER, Executive Director, Commission on Drug Safety